INVITATION TO MATHEMATICS

Presented by

Dennis Ragan

to the

**Kishwaukee College
Library**

Invitation to Mathematics

WILLIAM H. GLENN
Formerly Mathematics Supervisor
Pasadena City Schools, Pasadena, California

DONOVAN A. JOHNSON
Professor of Mathematics Education
University of Minnesota

DOVER PUBLICATIONS, INC.

NEW YORK

Published in Canada by General Publishing Company, Ltd., 30 Lesmill Road, Don Mills, Toronto,
Ontario.
Published in the United Kingdom by Constable
and Company, Ltd., 10 Orange Street, London WC 2.

This Dover edition, first published in 1973, is an
unabridged republication of the work originally
published in book form by Doubleday & Company,
Inc., in 1962.

International Standard Book Number: 0-486-22906-8
Library of Congress Catalog Card Number: 72-92757

Manufactured in the United States of America
Dover Publications, Inc.
180 Varick Street
New York, N.Y. 10014

Contents

PART I

PART I

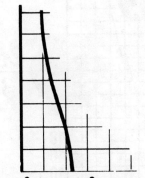

Invitation to
Mathematics

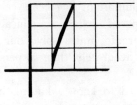

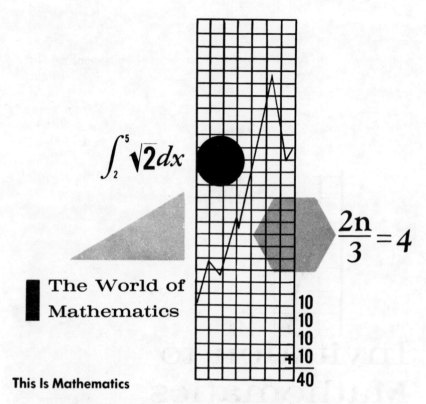

$$\int_2^5 \sqrt{2}\,dx$$

$$\frac{2n}{3} = 4$$

10
10
10
+ 10
40

The World of Mathematics

This Is Mathematics

Why has mathematics become so important in recent years? Why is our government spending millions of dollars to educate more mathematicians? Can the new electronic brains solve our mathematical problems faster and more accurately than a person and eliminate the need for mathematicians?

To answer these questions, we need to know what mathematics is and how it is used. Mathematics is much more than arithmetic, which is the science of numbers and computation. It is more than algebra, which is the language of symbols, operations, and relations. It is much more than geometry, which is the study of shapes, sizes, and spaces. It is more than statistics, which is the science of interpreting data and graphs. It is more than calculus, which is the study of change, limits, and infinity. Mathematics is all of these—and more.

Mathematics is a way of thinking, a way of reasoning. Mathematics can be used to determine whether or not an idea is true, or, at least, whether it is probably true. Mathematics is a field of exploration and invention, where new ideas are being discovered every day. It is a way of thinking that is used to solve all kinds of problems in the sciences, government, and industry. It is a language of symbols that is understood in all civilized nations of the world. It has even been suggested that mathematics would be the language that would be understood by the inhabitants of Mars (if there are any)! It is an art like music, with symmetry, pattern, and rhythm that can be very pleasing.

Mathematics has also been described as the study of patterns, where a pattern is any kind of regularity in form or idea. This study of patterns has been very important for science because pattern, regularity, and symmetry occur so often in nature. For example, light, sound, magnetism, electric currents, waves of the sea, the flight of a plane, the shape of a snowflake, and the mechanics of the atom all have patterns that can be classified by mathematics.

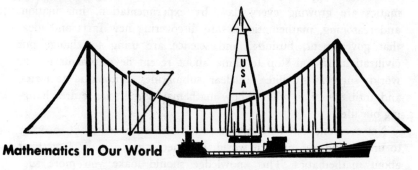

Mathematics In Our World

If we look back at the story of our civilization, we will find that mathematics has always played a major role. It has been the means of:

> measuring property boundaries
> predicting the seasons
> computing taxes
> navigating ships
> building homes and bridges
> drawing maps
> developing weapons and planning warfare
> understanding the motion of heavenly bodies
> increasing trade and commerce

During your lifetime, mathematics has been the tool for:

> discovering new scientific principles
>
> inventing new machines
>
> creating electronic brains
>
> developing strategy in games
>
> directing traffic and communications
>
> making new vaccines and medicines
>
> harnessing atomic energy
>
> navigating in space
>
> discovering new ores
>
> forecasting the weather
>
> predicting population growth

Both the applications of mathematics and the fields of mathematics are growing every day. By experimentation, imagination, and reasoning, mathematicians are discovering new facts and ideas that government, business, and science are using to change our civilization. If you stop to think about recent developments in our world, such as satellites, nuclear submarines, automatic factories, and antibiotics, you will see how mathematics and science are changing our life.

Not everyone can be a mathematician or a scientist, but in order to understand our modern world, it is necessary to know something about mathematics. This knowledge should make you more successful in school, at home, or in your future vocation. Just to be a good citizen in a nation when all these changes are taking place will require a knowledge of some mathematics. Certainly people in our government must be mathematically informed if they are to make wise decisions in our complex world of new ideas.

Of course, if you are interested in a career in science, statistics, or engineering, which are based on mathematics, you will need to become an expert in the field. Today there is a great demand for mathematicians to do research, teach, or find new applications of mathematics. The professional mathematician has often played an important role in the building of our civilization. The methods of reasoning used by the great mathematicians of the world and the products of their logic are more important than ever in our present culture.

What Does a Mathematician Do?

Mathematics is a tool for the map maker, the architect, the space navigator, the machine builder, the accountant, and many others. But accountants who work with problems of finance, or astronomers who measure the distance to Mars, or engineers who design a bridge, or scientists who invent a new plastic are usually not mathematicians in the strict sense of the word. True, they use many mathematical ideas that mathematicians have already discovered. But it is the job of the mathematician to discover new mathematics, to prove new ideas true, or to apply old mathematics to the solution of new problems. The mathematician doesn't enjoy adding a long column of numbers any more than you do. The mathematician is often concerned with interesting questions like this:

Suppose a five-year-old boy travels in space at the speed of light and returns to the earth in ten years. How old will the boy be when he returns?

According to Einstein's theory of relativity, the boy will be five years old when he returns! The boy does not grow old as long as he travels at the speed of light.

The ability of a mathematician to solve a problem depends to a great extent on his sensitivity to pattern. If he finds a certain striking pattern or regularity, he investigates it and tries to discover some meaning, some rule, some formula that will explain or describe the pattern. Thus, to be a good mathematician, you should enjoy the fascination of pattern. Pascal's triangle of numbers is a good illustration of the discovery of pattern by a mathematician. The French mathematician Blaise Pascal (1623-1662) studied the numbers obtained in mathematical relations like the following:

$$(a + b)^0 = 1$$
$$(a + b)^1 = 1a + 1b$$
$$(a + b)^2 = 1a^2 + 2ab + 1b^2$$
$$(a + b)^3 = 1a^3 + 3a^2b + 3ab^2 + 1b^3$$
$$(a + b)^4 = 1a^4 + 4a^3b + 6a^2b^2 + 4ab^3 + 1b^4$$

If you look at the numbers in gray, you will find the following pattern:

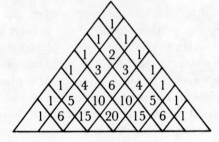

Figure 1

Notice that each row begins and ends in 1 but that the other numbers in the row are the sum of the two numbers above it. This number pattern is used to solve many problems in algebra and statistics.

It has been said that no one can be a mathematician without being somewhat of a poet. The mathematician likes to work with ideas. His work is largely thinking and reasoning. It is the kind of work that can be done while waiting for a bus, walking up a mountainside, or taking a bath. Whether it is done at an office desk or in a laboratory, it is exciting work and important for our nation and our world.

EXERCISE SET 1
Problems with Patterns

Solve these problems by finding a pattern that repeats regularly.

1. John Smith and his girl friend, Julie, both work. John is off duty every ninth day; his girl friend is off duty every sixth day. John is off duty today; Julie is off duty tomorrow. When (if ever) will they be off duty the same day?

2. Lay down 6 coins in a row, 3 heads up and 3 tails up, with one space between the two groups, like this: Ⓗ Ⓗ Ⓗ – Ⓣ Ⓣ Ⓣ

The problem is to change the positions of the coins by moving only one coin at a time to make this: Ⓣ Ⓣ Ⓣ – Ⓗ Ⓗ Ⓗ

A coin may jump over one adjacent coin into an empty space or it may be moved one space into an empty space. You may not move a coin backward! If you begin by using 2 coins instead of 6, then 4, you will soon discover the pattern.

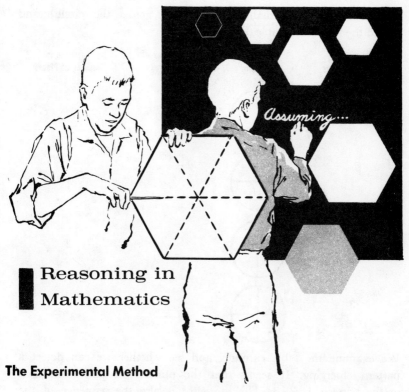

Reasoning in Mathematics

The Experimental Method

We often see mathematical problems involving coins, birthdays, holidays and the like, but mathematicians are not mainly interested in everyday problems. They are more concerned with using imagination, intuition, and reasoning to find new ideas and to solve puzzling problems. They enjoy exploring new ideas, trying different methods for solving problems, and stating new ideas in clear and concise language.

One method used by the mathematician in discovering new ideas is to perform experiments. This method is like the one used by a scientist in a laboratory. It is called the *experimental method* or *inductive reasoning*. Let's see how the experimental method can be used to solve a problem like this one:

> What is the maximum number of pieces obtained when you make cuts all the way across a circular cake, without cutting through the same point more than twice?

We can solve this problem experimentally. We don't need a cake, of course; a series of circles will do. We proceed to make various numbers of cuts (actually, lines drawn across the circle) and record the results in a table, like this:

Number of Cuts	Drawing	Number of Pieces	Increase in Number of Pieces
0		1	
1		2	1
2		4	2
3		7	3

We examine the table carefully and see whether we can detect a pattern emerging. It seems that the number of pieces increases in the pattern 1, 2, 3. This sequence is also the sequence of the increase in the number of cuts. Will this pattern of increase continue? Let us see whether it does or not by trying a few more cuts and recording our information, like this:

Number of Cuts	Drawing	Number of Pieces	Increase in Number of Pieces
4		11	4
5		16	5

Does the pattern continue? It does, for the increases now run 1, 2, 3, 4, 5. This pattern enables us to predict 22 pieces from 6 cuts, 29 pieces from 7 cuts, or the number of pieces obtained from *any* number of cuts. How many pieces would you predict for 8 cuts? 9 cuts?

Reasoning of this type, in which a general conclusion is made after considering specific examples, is called *inductive reasoning*.

Here is another simple experiment, which illustrates a well-known mathematical fact. Cut several triangles of different size out of paper. Tear off the corners of each one and fit them together as shown in Figure 2b below:

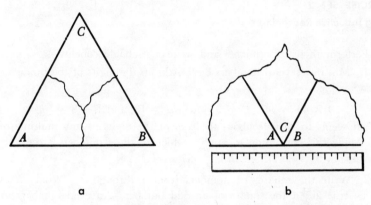

a   b

Figure 2

Do the angles of each triangle together form a straight line? If you place a ruler along the edge as shown in Figure 2b, it appears that the corners combine to form a straight line.

This experiment *suggests* that the measures of the angles of a triangle add up to a straight angle, or 180°. But no matter how many triangles we try out, we can never be sure that the angles of *every* triangle add up to 180°. Maybe some very oddly shaped triangle would not give this result. Hence, any conclusions we draw from experiments like this one are said to be *probably* true. Ideas discovered by this inductive or experimental method are *often* true but are not *always* or *necessarily* true.

Consider the following arithmetic computations:

$$2 \times 2 = 4 \qquad 2 + 2 = 4$$
$$\tfrac{3}{2} \times 3 = 4\tfrac{1}{2} \qquad \tfrac{3}{2} + 3 = 4\tfrac{1}{2}$$
$$\tfrac{4}{3} \times 4 = 5\tfrac{1}{3} \qquad \tfrac{4}{3} + 4 = 5\tfrac{1}{3}$$
$$\tfrac{5}{4} \times 5 = 6\tfrac{1}{4} \qquad \tfrac{5}{4} + 5 = 6\tfrac{1}{4}$$

From these examples, we might inductively conclude that we always obtain the same result when adding or multiplying the same two numbers. But you know that this conclusion is false, for you can disprove it with a simple example. This illustrates a serious shortcoming of inductive reasoning. By considering examples that are really special cases, it is possible to reach false conclusions. We can disprove a false conclusion by one example where the conclusion is not *true*.

EXERCISE SET 2
Using Inductive Reasoning

Perform these experiments and write a probable conclusion.

1. Multiply several numbers by 9. Add up the digits of the products; for example:

$$9 \times 43 = 387 \qquad 3 + 8 + 7 = 18, \text{ and } 1 + 8 = 9$$

What seems to be true about the sum of the digits of any multiple of 9? Apply this conclusion to predict which of these numbers are evenly divisible by 9: 477, 648, 8766

2. Write the squares of numbers from 1 through 20. What seems to be true about the square of an odd number? An even number? A number divisible by 5? What would you predict about the square of 22?

3. Measure the circumference and diameter of several round objects such as a coffee can, plate, lampshade, wastebasket, phonograph record, bicycle wheel. Divide the circumference by the diameter. Compare the quotients and state a conclusion obtained from your work.

4. Make a pendulum out of a piece of string and a weight. Make the pendulum 10 inches long. Count the number of times the pendulum swings back and forth in 10 seconds. Count the swings for pendulum lengths of 20, 30, and 40 inches. What conclusion can you state about pendulum swings related to the length of the pendulum?

Proof by Deductive Reasoning

Let's try another experiment with geometric figures. Draw several geometric figures with different numbers of sides, such as the ones shown in Figure 3.

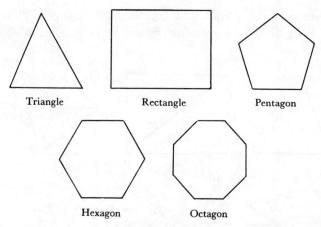

Triangle   Rectangle   Pentagon

Hexagon   Octagon

Figure 3

Measure the angles of these figures. You should get results like these:

Geometric Figures	Number of Sides (n)	Sum of the Measures of All Angles (S)	Number of Straight Angles (180°) in Each Sum
Triangle	3	180°	1
Rectangle	4	360°	2
Pentagon	5	540°	3
Hexagon	6	720°	4
Octagon	8	1080°	6

How does the number of sides of each figure compare with the number of straight angles (180°)? This table suggests the following conclusion: "The sum of the measures of the angles of a geometric figure equals the product of 180° multiplied by 2 less than the number of sides $(n - 2)$." This could be written as a formula, $S = (n - 2)\ 180°$.

This experiment is another example of the inductive method. When using this method it is necessary to repeat the measurements many times to see if the same pattern is always obtained. Of course, it is impossible to work with *all* possible geometric figures, and sometimes our measurements are not exact. This means that we can't be sure that our conclusion will apply to every geometric figure. Thus the inductive method gives us an answer that is only *probably* true.

However, we can arrive at a conclusion by using a different method. Let's divide the geometric figures into triangles, as shown in Figure 4.

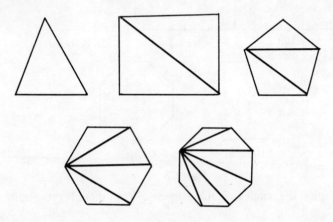

Figure 4

From the drawings in Figure 4, we make the assumption that a geometric figure can be divided into $(n-2)$ triangles, a number of triangles that is 2 less than the number of sides. From our previous experiment with triangles, we assume that the measures of the angles of a triangle add up to 180°. It is also assumed that the angles of any geometric figure will be made up of the sum of the angles of the triangles into which it can be divided. Then we can conclude that the measure of the angles of any geometric figure is $(n-2)$ 180°.

Let's examine the reasoning just used. We started with several ideas that we either assumed to be true or had previously established. Then we used these ideas to reach a conclusion by the force of reasoning. No measurements were made. By starting with an assumption about the number of triangles in any geometric figure, we found a specific conclusion about the angles of any geometric figure. Such a method of logical reasoning is called *deductive reasoning*. By deductive reasoning we obtain a specific conclusion from other ideas or assumptions.

Of course, the truth of our conclusion depends upon the truth of the starting assumptions and ideas. Here is another, very simple example of deductive reasoning. Suppose we assume every student in the ninth grade takes mathematics. If we know that Bill Jones is in the ninth grade, we can deduce that Bill Jones is taking mathematics. This specific conclusion is dependent upon the assumption with which we started our reasoning. Of course, we can be *certain* our conclusion is true only if our original assumption that all ninth graders take mathematics is true.

EXERCISE SET 3
Testing Your Reasoning Skill

Use experimentation or reasoning to answer these puzzles.

1. If 3 cats kill 3 rats in 3 minutes, how long will it take 100 cats to kill 100 rats?

2. I have two current United States coins in my hand. Together they total 55 cents. One is not a nickel. What are the two coins?

3. A bottle and a cork together cost 55 cents. The bottle costs 50 cents more than the cork. How much does the cork cost?

4. A monkey is at the bottom of a 30-foot well. Each day he jumps up 3 feet and slips back 2 feet. At that rate, when will the monkey be able to reach the top of the well?

Reasoning, Logic, and Proof in Mathematics

We have examined two types of reasoning used in mathematics. We have seen that both are useful, but that both also have drawbacks. Inductive reasoning is useful in making discoveries, but it can lead to false conclusions if the examples considered are not representative or are misinterpreted. Deduction will produce correct conclusions, but only if you start with correct assumptions. These two methods are often used together in mathematics: induction to develop acceptable assumptions, deduction to derive true conclusions from the assumptions.

Man's first experience with mathematics was of an inductive nature. The ancient Egyptians and Babylonians developed many mathematical ideas through observation and experimentation and made use of this mathematics in their daily life. Then the Greeks became interested in philosophy and logic and placed great emphasis on reasoning. They accepted a few basic mathematical assumptions and used them to prove deductively most of the geo-

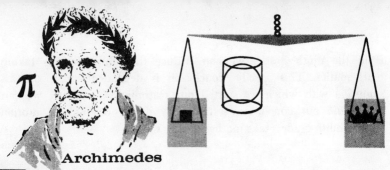

Archimedes

(287-212 B.C.)

Archimedes, the greatest Greek mathematician and scientist, was a counterpart of the modern-day absent-minded professor. When he was contemplating a problem, he forgot about food, rest, or comfort. He would sit for hours musing over some geometric figure which he had drawn in the ashes on the hearth. When a battle was raging over his city, Syracuse, a Greek colony in what is now Sicily, he was pondering over a geometrical figure he had drawn in the sand. A shadow fell over his diagram; and when he protested the interruption of his thoughts, a Roman soldier killed him. The Romans were not dreamers like the Greeks, but no Roman made contributions to mathematics or science such as those attributed to Archimedes.

Archimedes is famous for the many mechanical devices he invented. Many of these were military machines. He also discovered the principle that states that an object submerged in a fluid is buoyed up with a force equal to the weight of the fluid displaced. The story is told that he suddenly thought of the idea while taking a bath and ran to announce his discovery to the king without taking time to dress.

In mathematics, Archimedes is noted for his determination of the value of π. By comparing circles with polygons (many-sided figures) and increasing the number of sides of the polygons, he computed π to be a little more than $\frac{223}{71}$ and a little less than $\frac{220}{70}$. He proved many formulas for the area and volume of geometric figures, one of which compares the volume of a cylinder and of a sphere. One of his ideas was similar to the calculus invented by Newton.

Archimedes is supposed to have said, "If you give me a long enough lever, I can move the world." But he was more successful in "moving" the world forward in mathematics and science than he was in moving it physically.

metric facts we know today. Hence deductive proof became an important part of mathematics.

Since the time of the ancient Greeks, deduction has been the most important type of reasoning used in mathematics. However, mathematicians, like scientists, still discover new ideas from hunches, intuition, analogies, guesses, experiments. They then work out rigorous proofs to check the truth of the new ideas. This formal proof leaves nothing to the imagination. They use assumptions and definitions and previously proved statements to prove new statements. Usually they do not say, "Such and such is true." Instead they make statements like, "If A is true, then B is true." And they realize that the conclusion, B, depends upon the starting assumption, A, and might be true only in the world of mathematics with no apparent application or illustration in the physical world. For example, by using a logical proof, two Polish mathematicians, Stefan Banach and Alfred Tarski, have proved, from a mathematical standpoint, that a solid sphere the size of a pea could be divided into a finite number of pieces and then re-assembled into a sphere the size of the sun! No wonder mathematics is considered an unusual science.

EXERCISE SET 4
Recognizing and Using Types of Reasoning

1. Classify each of the following as examples of either inductive or deductive reasoning:

    a. After trying out several different objects and liquids, the ninth-grade science class concluded that an object will float in a liquid if it displaces a weight of liquid equal to its own weight.

    b. Only high school seniors can enter the mathematics contest. John is in the mathematics contest. John is a high school senior.

    c. Fifty men took a certain brand of vitamin pills for two months. Forty-five of the group gained weight during that period. We conclude that the vitamin pills cause people to gain weight.

2. The natives of a remote section of Africa are all members of either one of two tribes, Mau Mau and Bau Wau. To a stranger, they look exactly alike. But the members of the Bau Wau tribe always tell the truth, while those of Mau Mau blood always lie.

To this section came an explorer, who met three natives.

"Of what tribe are you?" the explorer asked the first.

"Chsz cinth cstrm," replied the native.

"What did he say?" asked the explorer of the second and third natives, both of whom spoke some English.

"He say he Bau Wau," said the second.

"He say he Mau Mau," said the third.

To what tribes did the second and third natives belong?

3. Three men who share a room at a hotel are charged $10 each, or $30 in all. The proprietor, after some reflection, decides that he has overcharged them, since they are sharing a room, so he gives a bellboy $5 to return to them. The bellboy, not being able to divide $5 into three equal parts, pockets $2 for himself and returns only $1 to each man. That makes the room cost each man $9, or $27 for all three. If we add this $27 to the $2 the bellboy keeps, we get a total of $29. Yet the men paid $30. Where is the other dollar?

4. A king who wishes to choose a prime minister decides to test the mentality of the three top candidates for the position. He tells the candidates that he will blindfold each one and then mark either a red cross or a blue cross on the forehead of each. He will then remove the blindfolds. Each candidate is to raise his hand if he sees a red cross and drop his hand when he figures out the color of his own cross. The king then blindfolds each candidate and proceeds to mark a red cross on each forehead. He then removes the blindfolds. After looking at each other, the prospective prime ministers all raise their hands. After a short time, one candidate lowers his hand and says, "My cross is red," and gives his reasons. Can you duplicate his reasoning?

5. The first customer in a grocery store one morning gave the grocer a $10 bill for $3 worth of groceries. The grocer, having no change, took the $10 bill across the street to the druggist, who gave the grocer 10 one-dollar bills. The grocer then gave his customer $3 worth of groceries and 7 one-dollar bills. About noon, the druggist brought the $10 bill back to the grocer, claiming it was counterfeit and demanding his money back. The grocer gave him 10 one-dollar bills and took back the counterfeit. How much did the grocer actually lose?

6. A chessboard has 64 squares. Suppose that you have 32 dominoes and each domino covers exactly two chessboard squares. The 32 dominoes may then be used to cover all 64 squares. Suppose that you cut off two squares, one at each diagonally opposite corner of the board. Discard one domino. Is it possible to place the 31 dominoes on the board so that the remaining 62 chessboard squares are covered? Show how it can be done or prove it is impossible.

The Fields of Mathematics

A Mathematical Structure

Mathematics started by answering these questions: "How many?" "When?" "How far?" "How big?" and "In what direction?" These questions are still being answered by mathematics. However, the questions proposed by new problems, such as those on travel in space or the energy of the atom, require new mathematical ideas. New numbers, new measurements, and new ways of finding relationships are needed. One of the most important new ideas in mathematics involves new ways of proving new facts. As a result of these needs, we now have more than eighty different branches of mathematics. In fact, more new mathematical ideas are being created every day than any one person could read in a day.

Each branch or field of mathematics has been developed and organized logically. For this reason, each field of mathematics is called a *logical structure* or *system*.

To build a mathematical system, the mathematician begins with a set of undefined words and a set of unproved assumptions. He must have words to express his ideas and assumptions from which he can deduce his ideas. He then develops rules that describe the things that can be done with the mathematical system. Next he invents symbols to simplify the use of the rules. These words, symbols, or rules do not need to have any reference to objects of the world around us. Mathematics often includes things that have never been seen or experienced. Some persons think that all mathematics has to be verified by concrete objects, measurements, or experimentation. This is not so, for mathematics belongs to the realm of ideas, imagination, and fantasy. But mathematics is unique in that it is also one of the most practical studies known to man.

Arithmetic: A Sample Mathematical Structure

Let's see how elementary arithmetic illustrates a deductive system. Arithmetic begins with the number 1 as an undefined term. With this unit all other numbers can be defined. The operations of addition and multiplication are also undefined terms. We describe these operations with addition and multiplication tables, but we do not give a definition of them. However, we define the operations of subtraction and division in terms of addition and multiplication, respectively.

Next, we state some assumptions about numbers that seem to agree with our experiences. For example, we assume the associative law to be true; that is,

$$2 + (3 + 4) = (2 + 3) + 4$$

We also assume that the sum of two numbers is always another number.

Then we define our numbers in terms of the unit 1 and addition:

$$2 = 1 + 1, \quad 3 = 2 + 1, \quad 4 = 3 + 1,$$

and so on.

These undefined terms, undefined operations, assumptions, and definitions are then used to prove new relationships called *theorems*. Let's see how we can use this deductive system to prove that $2 + 2 = 4$.

A Sample Logical Proof

Statements	Reasons
1. $2 + 2 = 2 + (1 + 1)$	**1.** 2 is defined as equal to $1 + 1$
2. $2 + (1 + 1) = (2 + 1) + 1$	**2.** Associative law
3. $(2 + 1) + 1 = 3 + 1$	**3.** 3 is defined as equal to $2 + 1$
4. $3 + 1 = 4$	**4.** By the definition of 4

In a similar manner, we can prove most relationships in arithmetic that we usually take for granted. This method of deduction can be applied to other mathematical systems and concepts. In each case, the mathematical structure developed is based on undefined terms, assumptions, definitions, and the proofs of theorems.

EXERCISE SET 5
Arithmetic Proofs

    1. Prove that $2 + 3 = 5$, as in the sample proof above.
    2. Assume the distributive law for multiplication to prove that $2 \times 2 = 4$. The distributive law may be illustrated as follows: $2(3 + 1) = 2 \times 3 + 2 \times 1$.

More About Arithmetic

    We have seen that arithmetic is a logical mathematical system. But arithmetic began as a language to answer questions about man's everyday life. To answer these questions, it was necessary to invent numbers and measures and a way of combining and comparing these numbers. This finally led to the development of the science of numbers or arithmetic.

    It took man a long time to invent a workable system of symbols for numbers. These symbols for numbers are called *numerals*. People worked with numerals for a long time before a zero symbol was invented. We know that our base ten numeration system, which uses the place value principle, is much better than many ancient systems of numeration. But our system is not the only system, nor is it necessarily the best one. Some people feel that a numeration system using twelve basic symbols would be superior to our base ten system. Another numeration system, called the *binary* system, which uses only two symbols, 0 and 1, has found use in new computers and other electronic devices.

Binary Numbers			
0	zero	100	four
1	one	101	five
10	two	110	six
11	three	111	seven

Gottfried Wilhelm Leibniz
(1646-1716)

Leibniz is best known as a mathematician, but he was a "master of all trades." He contributed to law, religion, politics, history, philosophy, and science, as well as mathematics.

Leibniz attended the university in his native city of Leipzig, in Germany, and by the age of seventeen he had his bachelor's degree. He would have had his doctor's degree at the age of twenty if the university faculty had not been jealous of the great knowledge of this brilliant young man. He spent most of the rest of his life as a kind of traveling diplomat. Many of his best ideas came to him while he was traveling over the rough roads of seventeenth-century Europe.

$$\int_{2}^{10} x \, dx$$

One of Leibniz's greatest interests was the development of a mathematics that would answer all questions in all fields. This led him to a discussion of logic which has been the basis of modern symbolic logic.

Leibniz was a religious man and wrote much about religion. Even his invention of binary numbers was related to his religious beliefs. He considered God as represented by 1, and the void, or nothing, by 0. Just as God could create all things out of the void, so could all numbers be represented in a binary system just by using 1 and 0.

Leibniz invented the calculus at about the same time as Newton, and there has been much controversy as to which man invented it first. Leibniz also invented a calculating machine that could add, subtract, multiply, divide, and extract roots.

Leibniz could probably have been the greatest of all mathematicians, but two things distracted him. One was his love for money, and the other was his curiosity about all fields of human knowledge. We can only wonder what mathematics he could have invented if he had devoted his whole time and attention to it.

The study of numbers has always been a fascinating topic. Even the most common numerical problem can suggest new questions or ideas. See if you can explain the method used in solving the following division problem:

$$
\begin{array}{r}
23\overline{)552} \\
230 = 10 \times 23 \\
\hline
322 \\
230 = 10 \times 23 \\
\hline
92 \\
92 = 4 \times 23 \\
\hline
24
\end{array}
$$

$$552 \div 23 = 10 + 10 + 4 = 24$$

We can often find unusual numeral patterns that give new insight into number relations. A good illustration is the pattern of multiples of 9.

$1 \times 9 = 9$	$0 + 9 = 9$
$2 \times 9 = 18$	$1 + 8 = 9$
$3 \times 9 = 27$	$2 + 7 = 9$
$4 \times 9 = 36$	$3 + 6 = 9$
$5 \times 9 = 45$	$4 + 5 = 9$
$6 \times 9 = 54$	$5 + 4 = 9$
$7 \times 9 = 63$	$6 + 3 = 9$
$8 \times 9 = 72$	$7 + 2 = 9$
$9 \times 9 = 81$	$8 + 1 = 9$
$10 \times 9 = 90$	$9 + 0 = 9$

Note how the tens digit goes up from 1 to 9 while the units digit goes down from 9 to 0. After 45, the products are the previous numerals written backwards.

There are many other interesting number patterns. For instance, triangular patterns can be used to represent numbers like 1, 3, 6, and 10.

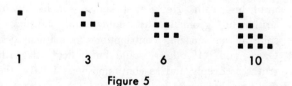

1   3   6   10

Figure 5

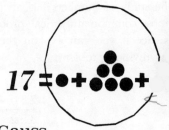

$$17 = \bullet + \begin{smallmatrix}\bullet\\\bullet\bullet\end{smallmatrix} + \begin{smallmatrix}\bullet\\\bullet\bullet\\\bullet\bullet\bullet\end{smallmatrix} +$$

Karl Friedrich Gauss
(1777-1855)

Karl Friedrich Gauss, together with Newton and Archimedes, is thought of as one of the three greatest mathematicians who have ever lived. Gauss was born in Germany in 1777, the son of poor parents. He was gifted in mathematics from his childhood. Gauss himself said he learned to reckon before he could talk! When he was ten years old, Gauss astonished his teacher with his mathematical talent by finding the sum of $81,297 + 81,495 + 81,693 + \ldots + 100,899$ in as little time as it took the teacher to state the problem. From that time on, he began his mastery and original thinking in mathematics. Arithmetic was Gauss' favorite field of study the rest of his life.

Fortunately, the Duke of Brunswick supplied money so that Karl could go to college at the age of fifteen. By the time he was eighteen, he had discovered new laws in the theory of numbers and invented a new statistical method called "least squares" that was used to determine a geometric figure that would best describe a set of data. He was quite excited when he discovered that every positive integer is the sum of three triangular numbers, for example, $17 = 1 + 6 + 10$. In the same year he also discovered how to construct a regular polygon with 17 sides.

Gauss became acquainted with many other mathematicians, but he considered Newton as his ideal. Although Gauss was a friendly individual, he showed contempt for anyone who pretended to know everything and would never admit a mistake. He lived his life in modest circumstances, continuing to make great contributions to mathematics until his death. He is best known for his discoveries in arithmetic, geometry, astronomy, and statistics. But despite all his amazing contributions to mathematics, modest Gauss said, "If others would but reflect on mathematical truths as deeply and as continuously as I have, they would make my discoveries."

Then there are patterns of squares that represent numbers like 1, 4, 9, 16, 25, 36, 49, 64, 81, 100.

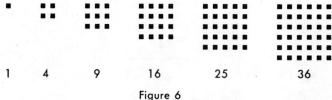

1 4 9 16 25 36

Figure 6

A strange relationship for square numbers is that they are equal to the sum of consecutive odd integers (whole numbers), like this:

$$4 = 1 + 3$$
$$9 = 1 + 3 + 5$$
$$16 = 1 + 3 + 5 + 7$$

Some numbers, like 6, 10, 15, can be represented with a rectangular pattern.

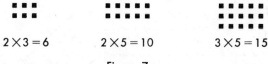

$2 \times 3 = 6$ $2 \times 5 = 10$ $3 \times 5 = 15$

Figure 7

Some numbers, such as 2, 3, 5, 7, 11, and 13, cannot be used to form either squares or rectangles. These numbers have no divisors except themselves and 1 and are called *prime numbers*. Mathematicians have been trying for many years to find relationships between prime numbers so that they could write a formula to describe these numbers and tell how to find them. No one has yet found a formula that will give all the prime numbers. Hence a good part of the work with prime numbers is still inductive.

Let's examine some interesting prime number patterns. If we use an arrow (⟶) to represent a prime number as a multiplier, and an asterisk (*) to represent a product, we can make drawings to show that every whole number is the product of prime numbers beginning with 1. For example, the diagram for 4 is

1 2 4

* ⟶ * ⟶ *

2 2

This diagram tells us that $1 \times 2 = 2$ and $2 \times 2 = 4$. The pattern for 7 is

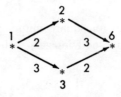

while 9 is

$$\begin{array}{ccccc} 1 & & 3 & & 9 \\ * & \xrightarrow{3} & * & \xrightarrow{3} & * \end{array}$$

But 6 can have two different paths, like this:

Note that the length of the arrows is not related to the size of the multiplier. Other patterns might look like this:

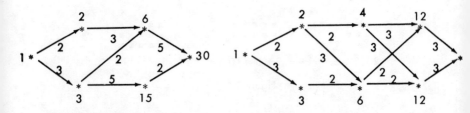

What kind of numbers have a straight line as a pattern? What kind of numbers have a complex figure as a pattern?

These examples illustrate how the patterns found in number relationships are of interest to mathematicians. The theory of numbers, a field of mathematics concerned with the study of number patterns, has been responsible for the discovery of many new mathematical ideas. We see that simple arithmetic can be the basis for many intriguing mathematical ideas and relationships.

EXERCISE SET 6
Some Arithmetic Puzzlers

1. There are between 50 and 60 eggs in a basket. If I count them out 3 at a time, I have 2 left over, but if I count them out 5 at a time, I have 4 left over. How many eggs are there in the basket?

2. The following is a problem in addition in which each letter represents a number and two different letters cannot be the same number:

$$
\begin{array}{r}
\text{S E N D} \\
+ \text{M O R E} \\
\hline
\text{M O N E Y}
\end{array}
$$

Can you figure out what numbers the letters represent?

3. Use arrows for prime multipliers to show the pattern for 18.

4. Try this problem about measurement:

"Six little puppies sitting on a mat;
One weighs more because he is so fat.
Can you separate him from the other five,
Which all have the same weight, dead or alive?
Put them on a balance, but only twice,
To find the one so fat and nice."

A New Arithmetic

Most of us make use of a clock or a watch many times every day. Our clocks are machines that measure time, and any measuring device is mathematical in nature. When we compute with hour numbers we often obtain unusual results. For example, if it is 8 o'clock and you add 7 hours, we usually say the result is 3 o'clock, not 15 o'clock. This means that the sum of 7 and 8 with clock numbers is 3. This new type of arithmetic obtained from a "clock-like" number system is called *modular arithmetic*. Sometimes modular arithmetic systems are called *finite number systems* because they have only a specific number of numbers in the system.

Another finite number system can be built from the days of the week. Let us think of Sunday as day number 1, Monday as 2, Tuesday 3, Wednesday 4, Thursday 5, Friday 6, and Saturday 7. This finite number system has no number larger than 7. The sum of 4 and 5 in this system is 2. (Five days beyond Wednesday is Monday.)

In the number system used on our clocks, the sum of 8 and 9 is written $8 + 9 = 5$ (mod 12) and is read, "8 plus 9 equals 5 (mod 12)." The expression "mod 12" means that the number system has only 12 numbers. In our weekday number system the sum of 5 and 6 would be written $5 + 6 = 4$ (mod 7).

Let's consider a modulo system which has only five numbers and five number symbols, 0, 1, 2, 3, and 4. This is called a "mod five" system. We will describe, but not define, addition and multiplication in this system. We then can define subtraction as the opposite (inverse) of addition and division as the opposite of multiplication, just as we do in regular arithmetic. We describe the operations in this system so that they never produce a number that is not in the system.

In technical language we say that the system is a *closed system,* since every addition, subtraction, multiplication, or division with mod five numbers will give an answer which is one of these original five numbers. In this system, we count 0, 1, 2, 3, 4, 0, 1, 2, 3, 4, 0, 1, and so on, as we would count on a clock dial.

One way to learn the meaning of addition in the mod five system is to make a dial like the one in Figure 8 below. Make this dial and pointer out of cardboard. Attach the pointer to the dial with an eyelet or a paper staple. This dial is a scale for a mod five number scale, just as the number line in Figure 9 is a scale for our ordinary numbers.

Mod Five Number Scale

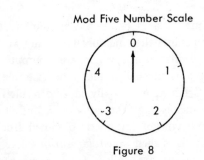

Figure 8

Scale for Our Number System

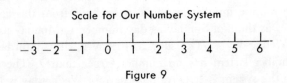

Figure 9

On our regular number scale, we add by counting spaces to the right. In the mod five system, we add by counting clockwise. To add 3 and 4 in mod five, place the pointer at 3, then move the pointer 4 spaces clockwise. You will end at 2, so 3 + 4 = 2 (mod 5).

We see that addition in this mod five system is described in the same way as ordinary addition, except when the sum is larger than 4. When the sum is larger than 4, we divide the sum by 5, discard the quotient, and use only the remainder in place of the ordinary sum. For example,

$$3 + 4 = 7, \quad 7 \div 5 = 1, r \, 2,$$

so

$$3 + 4 = 2 \pmod 5.$$

Likewise,

$$2 + 3 = 0 \pmod 5$$

and

$$4 + 4 + 4 = 12, \quad 12 \div 5 = 2, r \, 2,$$

so

$$4 + 4 + 4 = 2 \pmod 5.$$

Isn't this a strange number system? And yet it is the kind of number system we use for all machines with dials or with a finite number of numbers. Scientists have also found it useful in the study of atoms which have shells of electrons. It is really simple because there are so few numbers in the entire system. There are no negative numbers and no fractions, and yet you can solve equations with these numbers. And you can't say that one number is more than or less than another number.

Does 2 marbles plus 2 marbles always equal 4 marbles? Of course it does. But sometimes we use different meanings for numerals such as 2 and 4. And nobody knows whether we will ever discover new numbers that will be better than the numbers we now have. Perhaps you will be the one to find new numbers that will lead to important mathematical discoveries.

EXERCISE SET 7
Modular Arithmetic Problems

1. Use the mod five dial to find the sums:

   a. $4 + 4 =$ e. 3   f. 2

   b. $3 + 2 =$ 4   1

   c. $3 + 3 =$ $\underline{+\,2}$ 3

   d. $2 + 4 =$ $\underline{+\,4}$

2. Does $3 + 4 = 4 + 3$ in the mod five system?

3. Complete this addition table for the mod five system.

+	0	1	2	3	4
0	0				
1		2			
2			4		
3				1	
4					3

4. Answer these questions with mod five numbers.

      a. Is the sum of two even numbers always even?

      b. Is the sum of two odd numbers always even?

      c. Is the sum of an odd number and an even number always odd?

5. Copy and complete the following mod five multiplication table:

×	0	1	2	3	4
0	0	0	0	0	0
1	0	1		3	4
2	0	2	4		3
3	0		1		
4	0	4			

6. In ordinary arithmetic, the question "7 minus 3 is what?" can be interpreted as "What, when added to 3, gives 7?" We say that subtraction is defined as the *inverse* of addition. Use this definition of subtraction to do the following problems:

      a. $4 - 3$ (mod 5)   d. $3 - 3$ (mod 5)

      b. $3 - 4$ (mod 5)   e. $2 - 0$ (mod 5)

      c. $2 - 4$ (mod 5)   f. $0 - 2$ (mod 5)

7. Make use of the definition of division as the inverse of multiplication to do the following mod five division problems:

      a. $\frac{4}{2}$ (mod 5)   d. $\frac{1}{2}$ (mod 5)

      b. $\frac{2}{1}$ (mod 5)   e. $\frac{3}{4}$ (mod 5)

      c. $\frac{2}{4}$ (mod 5)   f. $\frac{3}{2}$ (mod 5)

8. Devise a mod four number system and construct addition and multiplication tables for the system. Make up some subtraction and division problems in mod four and use your tables to solve them.

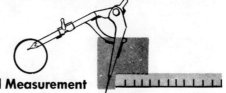

Geometry: Space, Shape, and Measurement

The mathematical study of space, shape, and measurement is known as *geometry*. Geometry tells us how to draw different shapes or figures and tells us many facts about the relationships among these figures.

The study of geometry has been basic for artists, engineers, and architects. Drawing accurate plans for a building, determining the effect of a strong wind on an airplane, painting a picture that has a balanced and pleasing design, all are related to the ideas about points, lines, angles, and shapes that are studied in geometry.

For centuries, man was curious about geometric mysteries such as the relationship between the diameter and circumference of a circle, whether parallel lines ever meet, whether space is curved and without a boundary. Even such a simple idea as a line has been interesting. You see, no one has ever seen a line. In geometry, a line has no width—only length. And no one knows if it is possible for a line to be "straight."

New Geometries

The geometry with which we are most familiar is called *Euclidean geometry*. This name is a tribute to the Greek scholar Euclid (323-285 B.C.), who collected, systematized, and recorded the geometric knowledge of his day. However, Euclidean geometry is not the only possible kind of geometry. By starting with some assumptions about lines that differ slightly from the assumptions of Euclid, mathematicians in the last century developed some strange geometries, all called *non-Euclidean geometries*.

One non-Euclidean geometry says that the shortest distance between two points is not a straight line, but, rather, a curved line. Does this sound strange? It does if we are talking about flat surfaces. But this idea makes sense if we consider a sphere or a ball instead of a flat surface. Indeed, navigators of ships and airplanes know that the shortest distance between two points on the surface of the earth (approximately a sphere) is a curve which we call a *great circle*.

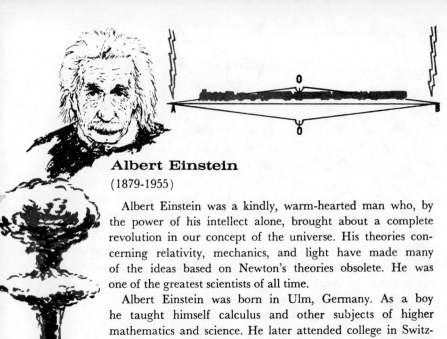

Albert Einstein

(1879-1955)

Albert Einstein was a kindly, warm-hearted man who, by the power of his intellect alone, brought about a complete revolution in our concept of the universe. His theories concerning relativity, mechanics, and light have made many of the ideas based on Newton's theories obsolete. He was one of the greatest scientists of all time.

Albert Einstein was born in Ulm, Germany. As a boy he taught himself calculus and other subjects of higher mathematics and science. He later attended college in Switzerland and, in 1902, secured a position in the Swiss Patent Office in Berne. While there, he had plenty of time to think, to dream, and to write papers on physics and mathematics. In 1905, he introduced the famous formula $E=mc^2$ in his first published article on the theory of relativity. This theory led to the discovery of nuclear energy. The formula predicted how much energy would be released by a nuclear reaction similar to that of an atomic bomb explosion.

Einstein knew he had important contributions to make to the world of mathematics and science, and he loved to work alone on new ideas. His dream was to find a mathematical formula that would unify all the forces of nature such as gravity, electricity, and atomic energy. His great discoveries were based on mathematical reasoning, not on experiments, and they illustrated the close relationship between mathematics and science. Einstein made important use of ideas from non-Euclidean geometry in his thinking. His ideas about the fourth dimension will become very useful when man travels into space.

Einstein spent his later years in the United States. Since he had experienced hardships because of religious prejudice when he lived in Germany, he was always ready to help anyone in trouble. His greatest worry was that we would misuse the power which he had helped to release through the mastery of the forces of nature.

Figure 10

With this idea of great circles, we can go a step further and reach another astonishing conclusion of non-Euclidean geometry— that the measures of the angles of a triangle drawn on a sphere may add up to more than 180 degrees. We can see this from the triangle in Figure 11.

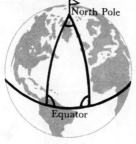

Figure 11

Suppose that the sphere is the earth. The base of the triangle is the equator. The two lines intersecting the equator are meridians, meeting at the North Pole and forming the triangle. The sum of the measures of the angles of this triangle is equal to the two 90° angles (spherical angles are measured by the degrees in their arcs) plus the angle formed at the North Pole.

Another kind of new geometry is one called *topology*. Unlike the geometry we usually study, topology is not interested in size or shape. Topology deals with lines, points, and figures, but these elements are allowed to change in size and shape in topology. Sometimes topology is called the *rubber-sheet geometry* because topological figures can be stretched and twisted, as they might be on a rubber sheet, and still remain the same, topologically speaking.

Topology is concerned with properties of position that are not related to size or shape. In the world of topology, a straight line like

AB below is the same as the curved line *CD* because each of them represents a path between two points.

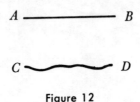

Figure 12

Topology also says that a triangle, a square, and a circle are the same. They all have one inside and one outside, and to go from the inside to the outside you must cross one and only one line.

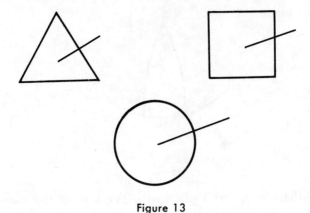

Figure 13

These new ideas of topology have been the means of solving many problems in mathematics.

EXERCISE SET 8
Some Different Geometry Problems

1. An explorer walks one mile south, turns and walks one mile east, turns again and walks one mile due north. He finds himself back where he started. At what locations on the earth's surface is this possible? (There is more than one location where this is possible.)

2. The blocks of this drawing are identically lettered. Roll them around with your imagination or sketch a pattern so that you can tell the letter that is on the side opposite Y. Opposite G. Opposite W.

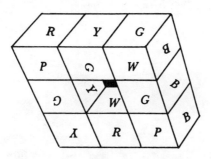

Finding Unknown Parts with Trigonometry

Trigonometry is another branch of mathematics that involves measurement, space, and shape. Trigonometry started from a study of the relationships of the sides of right triangles. Later it was extended to a variety of triangles. The relationships between the angles and sides of a triangle become the means of measuring distances indirectly, such as the height of a cliff or the distance to the moon. From Figure 14 we can get an idea of how the distance to the moon can be found. We know the diameter of the earth, AB, is equal to about 8,000 miles. This fact, plus the measures of angles A and B, gives us enough information to compute the distance to the moon, AM, by trigonometry.

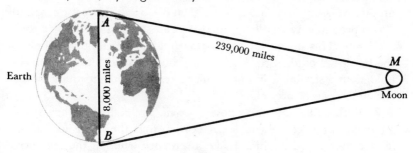

Figure 14

Trigonometry now goes far beyond this kind of problem. Trigonometry has established relationships and formulas that are very useful in the study of patterns such as those found in electricity, sound, and atomic energy.

$$5+3=8$$

$$?+3=8$$

$$x+3=8$$

$$x+y=z$$

$$z+y=8$$

Algebra, the Language of Mathematics

One of the things about mathematics that makes it such a powerful science is its use of symbols such as $+$ or 5. These symbols make mathematics a shorthand language. It is much simpler to write $5 + 3 = 8$ than to write "the sum of five and three is eight." Such symbols make it easy to think about mathematical ideas, and make it possible to show that ideas or relationships which are true for specific numbers may also be true for all numbers. For example, we know that if $5 + 3 = 8$ then $8 - 3 = 5$. If we use a, b, and c to represent any numbers, then we can say that if $a + b = c$ then $c - b = a$. This use of letters as placeholders for numbers is one of the big ideas of algebra. You have probably used letters as placeholders in a similar way in formulas like $c = \pi d$, the formula for the circumference of a circle.

Formulas like the one above, $c = \pi d$, are called mathematical sentences. In algebra you learn how to find the value for placeholders like c and d which make a sentence true. In the sentence, $x + 3 = 8$, what number can replace x so that this is a true sentence?

When a scientist performs an experiment to discover new facts, he often arranges his measurements in the form of a table. Then he tries to write a mathematical sentence like a formula that shows how the numbers in his table are related.

Let's look at the following set of data to see whether it can be expressed by a formula.

Measurement a	Measurement b	a + b	a×b
1	12	13	12
2	6	8	12
3	4	7	12
4	3	7	12
6	2	8	12
8	$1\frac{1}{2}$	$9\frac{1}{2}$	12
10	$1\frac{1}{5}$	$11\frac{1}{5}$	12
12	1	13	12

If we compare the measurements by addition $(a + b)$, we do not find a pattern for the sums. But if we look at the products of the measurements $(a \times b)$, we find that they are always 12. This tells us that the formula for this pattern is $a \times b = 12$. In this way, algebraic ideas are used to discover and state relationships.

A famous French mathematician by the name of René Descartes (1596-1650) showed how algebra and geometry are related. He drew a line to show that any number could be represented by a point on the line and that any point on the line could represent a number. For example, in Figure 15, point x represents 3 and point y represents 1.43786.

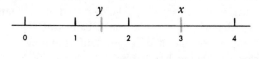

Figure 15

With two number lines drawn perpendicular to each other, we can use a point to represent a pair of numbers. For example, point x in Figure 16 represents the number pair $a=3$, $b=4$.

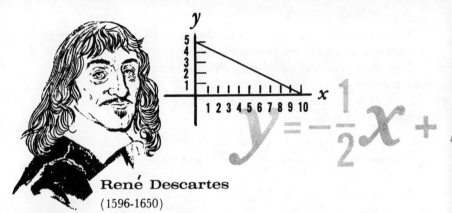

$$y = -\frac{1}{2}x + $$

René Descartes
(1596-1650)

René Descartes, the great French mathematician and philosopher, lived in a time of change, like our world of today. His century was a time of new ideas in science, mathematics, philosophy, and literature which were produced by such intellectual giants as Newton, Galileo, Fermat, and Shakespeare.

Descartes was very frail as a youngster and was permitted to spend much time in bed, where he studied and thought about the world. At the age of eighteen, though, he enlisted in the army. Fortunately, he was able to spend a great deal of time as a soldier thinking about important problems in mathematics and philosophy.

In November, 1619, while he was camped in a little village on the Danube, a great idea came to him in a dream. In the dream, he saw how algebra could be applied to geometry. This thought opened up a completely new field of mathematics called *analytic geometry,* which has been one of the most useful ways that scientists have of studying natural phenomena.

Fortunately, Descartes survived his army career and then roamed about Europe. At first he wasted much time in pleasure, but later he settled in Holland and devoted himself to the study of science, theology, and mathematics.

In the winter of 1650, he became the teacher of young Queen Christina of Sweden. Unfortunately the queen insisted on having her lessons at five o'clock in the morning, in her unheated library. Descartes was too frail to withstand the loss of sleep and the extreme cold of the Swedish winter and very soon fell ill of pneumonia and died. Had not the foolish young queen made such strange demands, who knows what further great ideas Descartes might have been able to contribute to the world?

If we locate points for the number pairs for *a* and *b* given in the table and join the points with a curved line, we get a part of a geometric figure called the *hyperbola*, as shown in Figure 16.

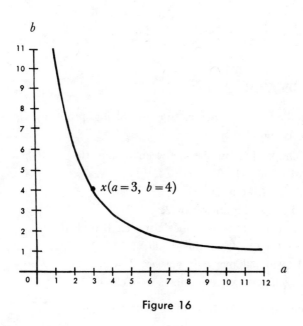

Figure 16

Thus the mathematical sentence, $a \times b = 12$, can be represented by a geometric figure.

EXERCISE SET 9
Picturing Algebraic Relationships

1. On graph paper, draw the geometric figure for the data for the equation $x^2 + y^2 = 25$.

Measurement x	Measurement y	x^2	y^2
0	5	0	25
1	4.9	1	24 (app.)
2	4.6	4	21 (app.)
3	4	9	16
4	3	16	9
5	0	25	0

2. Here is data for an object falling to the earth:

Time in seconds (t)	Distance fallen (S)	t^2
1	16	1
2	64	4
3	144	9
4	256	16

What formula shows the relationship of S and t? $S = ?$

Some Magic with Algebra

Algebra is best known for its use in solving problems. This is illustrated by this number trick:

> Think of a number between 0 and 10.
> Multiply it by 5.
> Add 7 to your answer.
> Multiply this answer by 2.
> Add any other number between 0 and 10.
> Subtract 3 from this result.
> If you tell me your answer, I will
>   tell you what numbers you chose.

Here is how this trick works with a friend.

Instructions	Your friend's arithmetic	Your solution with algebra
Think of a number between 0 and 10.	Thinks of 6.	Use x as a place-holder for this unknown number.
Multiply it by 5.	$5 \times 6 = 30$.	$5x$
Add 7 to your answer.	$30 + 7 = 37$.	$5x + 7$
Multiply this answer by 2.	$37 \times 2 = 74$.	$10x + 14$
Add any other number between 0 and 10.	Picks 8; $74 + 8 = 82$.	Use y as a place-holder. $10x + 14 + y$
Subtract 3 from this result.	$82 - 3 = 79$.	$10x + (14 - 3) + y$ $= 10x + 11 + y$
If you tell me your answer, I will tell you your numbers.	79	6 and 8

How can you get the unknown numbers 6 and 8? You set your algebraic expression equal to the number stated by your friend:

$$10x + 11 + y = 79$$

Then subtract 11 from your expression and your friend's answer to get

$$10x + y = 68.$$

Now you know that $10x$ is 10 times the number he chose first, or the tens digit 6, and the symbol y represents the number he added to the tens digit, or 8. Try this trick with your friends to demonstrate the magic you can do with algebra.

Probability, the Science of Chance

One of the most intriguing fields of mathematics is the study of probability, or chance. Probability tells us how often we may expect some event to occur. It is the field of study which is giving the scientist and the social scientist the tools to deal with our uncertain world. Computing the probability that something will happen is like looking into the future.

Probability is used in business, in government, and in science to make predictions. Using probability, an insurance company predicts how many homes are likely to burn every year; the federal government predicts how much income tax will be collected next year; and a scientist predicts the performance of a space ship before it takes off. Since we all run the risk of accidents, difficulties, or success every day, it is important that we know something about the mathematics of probability.

A simple illustration of probability is the tossing of a coin. If you toss a penny, it is just as likely to turn up heads as tails. When we toss several coins, we expect about half of them to turn up heads and the other half tails. So we say your chance of getting a head when you toss one coin is $\frac{1}{2}$. In more technical language, we say that the probability of an event happening is equal to the ratio of favorable ways to the total number of ways it could happen.

To see how this works with several coins, we need to study all the possibilities. If we toss 2 coins, the coins may fall four ways, as shown on the next page.

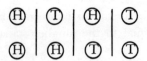

Since one of the four ways gives us 2 heads, we say the chance of tossing 2 heads with 2 coins is $\frac{1}{4}$.

If we toss 3 coins, the coins may fall eight ways, like this:

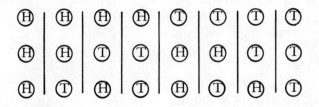

What are the chances of getting 3 heads? What are the chances of getting 2 heads and 1 tail?

EXERCISE SET 10
Problems in Probability

1. Suppose you have two dice. Make a table of all the different ways these dice might fall. What is the probability of turning up a total of 7 dots on the two dice?

2. Imagine that you have three boxes, one containing two black ties, one containing two white ties, and the third, one black tie and one white tie. The boxes are labeled for their contents—BB, WW, and BW —but someone has switched the labels so that every box is now incorrectly labeled. You are allowed to take one tie at a time out of any box, without looking inside, and by this process of sampling you are to determine the contents of all three boxes. What is the smallest number of drawings needed to do this?

Statistics: Making Sense Out of Data

Number facts called *data* or *numerical statistics* are found in every daily newspaper. In industry, on the farm, in state and national legislatures, at home, in school, and at church, statistics plays an important role. Information about the weather, sports, jobs, wages, prices, and population changes is daily news. These data are the basis for important decisions. If these decisions are to be the right decisions, the data must be read and interpreted intelligently.

An important application of statistics is the use of *sampling*. Sampling is used in public opinion polls, TV program popularity ratings, and quality control in industry. In sampling, we choose a small number of items which we think are typical of the whole and examine the sample. We can then say, if our sample has been properly chosen, that what is true of the sample is also true of the whole.

Here is a simple example of how sampling works. Suppose you wish to find out something about the size of the pebbles on a beach. Obviously, you cannot measure all the pebbles. Instead, you might select a sample of, say, 500 pebbles from all parts of the beach. The statistician has determined that 99 times out of 100 only 1 percent of all the pebbles on this beach will be larger than the largest pebble in the sample, so you can get a good picture of the size of all the pebbles from those in your sample. This same type of result is expected whether the sample is of people, fish, flowers, or numbers drawn out of a hat. An application of this method in industry might be to test samples of tires to determine the quality of all the tires produced. By testing a proper sample of 65 tires, the manufacturer can conclude that 90 percent of all the tires in the sample will almost certainly have a longer wearing life than the second tire to wear out during the test. By the way, the statistician calls 99 chances out of 100 "almost certain."

Statistical methods, such as the presentation of data in tables and graphs, help us to read and interpret data. Statistics also tells us how to compute measures such as averages or deviations that tell us the trend of the data. Relationships are determined by finding a measure of association called the *correlation coefficient*. Then probability is used to state the meaning of the data or to predict a result in the future. For example, a statistician may look at your mathematics test scores, tell you how you compare with an average student, predict your success in college, and tell you the probability that his prediction is right.

EXERCISE SET 11
Drawing Conclusions from Data

1. Which one of the following would be the best sample to use in deciding what car is most popular in your town?

    a. A sample of the owners of Chevrolets.

    b. A sample of the women drivers in your town.

    c. A sample of the people who do not own an automobile.

    d. A sample of the people who bought a car during the last year.

    e. A sample of everyone in your town.

2. Why aren't these samples good ones for finding out how much time the students in your school spend watching TV programs?

    a. The students in a mathematics class.

    b. The members of the band.

    c. The students at the football game.

    d. The students with lockers on the third floor.

    e. The students who ride the school bus to school.

3. What relationship is assumed in making these statements?

    a. Tarzan is the best soap because it floats.

    b. Ripple is the best bicycle to buy because it is the cheapest.

    c. If you eat Bumpers cereal, you will be a good athlete just like your football captain who eats Bumpers.

4. Collect data from your newspaper or school or friends to find answers to questions like these:

    a. How many hours do your classmates spend watching TV each week?

    b. How has the average score in basketball games changed for your school in the past 5 years?

    c. What is the average height of your friends?

    d. Do the students in your school who drive cars get "A" grades in mathematics?

    e. How many of your school's graduates go to college each year?

Infinity, Limits, Changing Quantities, and Calculus

How far is it to the end of space? How long ago did time begin? How many numbers are there in our number system? How many points are there on a line? The answer to all these questions is "an infinite number." By this, we mean that the number of items in the answer cannot be counted.

Infinity is an idea that is hard to understand and impossible to visualize. But mathematicians have invented ways of dealing with

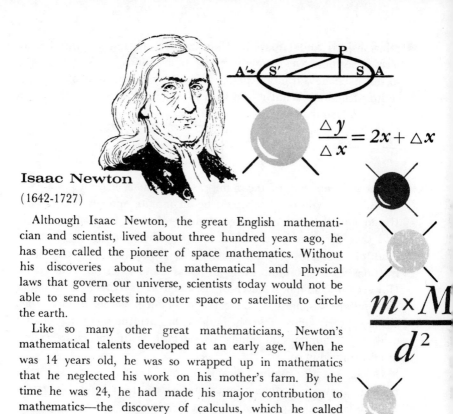

$$\frac{\triangle y}{\triangle x} = 2x + \triangle x$$

$$\frac{m \times M}{d^2}$$

Isaac Newton
(1642-1727)

Although Isaac Newton, the great English mathematician and scientist, lived about three hundred years ago, he has been called the pioneer of space mathematics. Without his discoveries about the mathematical and physical laws that govern our universe, scientists today would not be able to send rockets into outer space or satellites to circle the earth.

Like so many other great mathematicians, Newton's mathematical talents developed at an early age. When he was 14 years old, he was so wrapped up in mathematics that he neglected his work on his mother's farm. By the time he was 24, he had made his major contribution to mathematics—the discovery of calculus, which he called "fluxions." Even though he made this great discovery, his theories were still not fully developed, and it took him twenty years to solve a certain problem in calculus that he needed to prepare his most important scientific work. Today, a college student would find the same problem in his calculus textbook and could probably solve it in about half an hour.

Newton's fame as a mathematician was widespread. The story is told that the scientist John Bernoulli posed two very difficult problems in mathematics and gave mathematicians six months to solve them. A day after Newton received the problems he had solved them. Even in his old age, Newton's mathematical skill had not left him. When he was 74 years old, he accepted a challenge from Leibniz to solve a difficult problem and worked it out in one evening.

Isaac Newton was one of the great intellects of all time. He was called "an ornament of the human race." But, as great as he was, many of his ideas have been challenged today and are being modified by the work of scientists of our own time.

relationships and situations that involve an infinite number of terms. Suppose we have the series of fractions, $\frac{1}{2}$, $\frac{1}{4}$, $\frac{1}{8}$, $\frac{1}{16}$, and so on, each new fraction being formed by taking half of the preceding fraction. What is the sum of all the fractions of this form that you could write?

$$\frac{1}{2} + \frac{1}{4} + \frac{1}{8} + \frac{1}{16} + \cdots$$

$\frac{1}{2}$	$\frac{1}{4}$	$\frac{1}{8}$	$\frac{1}{16}$

There is no end to the number of fractions like this that you could write. We say there would be an *infinite* number of fractions. If you find the sum of a large number of these fractions, you will find that the sum is very close to 1. Mathematicians say that the *limit* of the sum of these fractions is 1 because you can write fractions like this until the sum is as close to 1 as you desire. However, the sum will never be as much as 1. Whenever we work with infinite amounts of things such as time, points, or numbers, we use special kinds of mathematics. One field of mathematics that deals with infinity and limits is *calculus*, a mathematics course usually studied in college.

Calculus is the mathematics that enables us to study the relationship between changing quantities. Here is a typical question that has been answered by calculus:

If a pebble is dropped from a cliff, at what speed would the pebble be traveling in 6 seconds?

In this problem, the distance changes as the time changes, and the speed also will change as the time changes. The methods of calculus have been used to solve this problem by finding a *limiting* value for the speed as the time is divided into smaller and smaller units. These methods would finally tell us that the speed at any instant would be equal to 32 times the number of seconds traveled. In 6 seconds, the pebble would be traveling at a speed of 192 feet per second.

EXERCISE SET 12
Limits and Changing Quantities

1. At exactly 12 o'clock, 2 bacteria are placed in a growing medium. One minute later there are 4 bacteria, in another minute they have increased to 8, in another to 16, and so on. At exactly 1 o'clock,

the growing mass of bacteria measures 1 gallon. At what time was there 1 quart of bacteria?

2. What is the limit of the unending decimal 1.99999...?

3. How many fractions are there between zero and one?

4. What is the limit of $1 - \frac{1}{2} - \frac{1}{4} - \frac{1}{8} - \frac{1}{16} - \ldots$?

5. Find examples of some numbers or measurements that are unending.

Sets: A Useful Mathematical Idea

One of the greatest new ideas in mathematics developed in the past century is *set theory*, invented by the German mathematician Georg Cantor (1845-1918). It has been a means of finding new facts and of proving old facts.

The idea of a set is very simple. A set is a collection of objects, numbers, persons, or ideas. You are already familiar with sets such as a set of books, a set of dishes, or a set of tools. You are even a member of a set of people such as your mathematics class, your scout troop, or your family.

Set ideas are used in many fields of mathematics. In arithmetic, we talk about sets of numbers. For example, the set of prime numbers less than 10 is { 2, 3, 5, 7 }.

In algebra, we talk about the solution sets for sentences. For example, the solution set for $x - 5 = 7$ is the number 12.

In geometry, we talk about the set of points that meet certain conditions. The set of points on line AB and also on the boundary of the circle is the points P and Q.

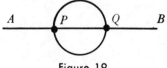

Figure 19

In statistics, we talk about sets of data. For example, the set of scores on a weekly math quiz was { 7, 10, 5, 6, 9, 7, 8, 4, 6, 8, 5, 7 }.

In probability we talk about the set of all possible events. For example, the set of ways three letters, *A,B,C,* can be arranged is { *ABC, ACB, BAC, BCA, CAB, CBA* }.

In measurement, we talk about a set of units of measure. A set of metric units of length is {kilometer, hectometer, decameter, meter, decimeter, centimeter, millimeter}.

In everyday situations, we talk about sets of persons or objects with certain characteristics. The set of all students in the algebra class and also on the basketball team, for example, might be { John, Bill, Steve }.

Set theory has been the basis for a new philosophy about mathematics. It has freed mathematics from working with individual numbers, permitting it to consider sets of numbers. Set ideas have also made it possible to solve many problems that involve relationships rather than numbers. And sets have been proved useful in setting up problems for electronic computers.

EXERCISE SET 13
Reasoning Problems Involving Sets of People

1. Mr. Jones and Mr. Smith have the same amount of money. Mr. Jones, however, has more money than Mr. Brown, and Mr. Brown has more money than Mr. Robinson. Another man, Mr. Carr, who has less money than Mr. Jones, but more money than Mr. Robinson, does not have as much money as Mr. Brown. Mr. Smith has less money than his friend, Mr. Stevens.

   a. Is Mr. Robinson richer or poorer than Mr. Smith?
   b. Is Mr. Carr richer or poorer than Mr. Jones?
   c. Is Mr. Stevens richer or poorer than Mr. Brown?
   d. Is Mr. Jones richer or poorer than Mr. Robinson?
   e. Is Mr. Brown richer or poorer than Mr. Carr?

If the difference between the amount of money owned by each man is $1250, and if the poorest man has $5, how much does each man have?

2. A girl once said: "The man I marry will be tall, not dark, quite heavy, will wear glasses, will carry a cane, and will be an American."

John is tall, fair, an American, and wears glasses, but does not carry a cane.

Richard is not short, wears glasses, carries a cane, is not fair, is not heavy, and is an American.

Ferdinand carries a cane, is dark and not too heavy, is not short and is certainly not fair. He wears glasses.

Which one of the three men will the girl marry?

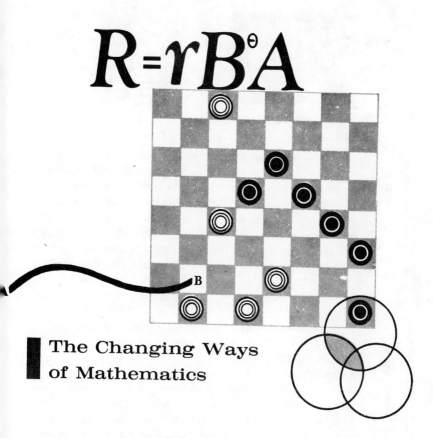

$$R = rB^{\theta}A$$

The Changing Ways of Mathematics

Developments in New Mathematics

Although mathematics is one of the oldest sciences, it is growing like a boy in high school. New ideas, new vocabulary, new methods, new problems are constantly being created. Mathematics has never been changing as rapidly as it is today.

One of the most exciting new fields of mathematics is *game theory*. The mathematician John von Neumann contributed a great deal to this theory. Game theory is a mathematical way of describing and analyzing competition among groups of people such as clubs, athletes, or checker players. Game theory gives a mathematical analysis of what decisions or moves should be made to "win" in a competitive situation. Thus, game theory can be used to make decisions in business, in foreign affairs, and in the detection of a murderer.

John von Neumann
(1903-1957)

John von Neumann was one of the outstanding mathematicians of our day. Like so many other mathematicians, he made important contributions to both science and mathematics. Von Neumann was particularly fascinated by games of strategy and chance. So it is not surprising that he was one of the persons who opened the new field of mathematics called *game theory.* Using the probabilities involved in games of chance and working with strategies that produce "winners" in decision-making games, von Neumann's theory of games can solve problems in economics, science, and military strategy.

Von Neumann was born in Budapest, Hungary. When he was 6 years old, he was able to do division problems like 78,463,215÷49,673,235 in his head. By the age of 8, he had mastered college calculus and as a trick could memorize on sight the names, addresses, and telephone numbers in a column of a telephone book. When he was only 23 years old he wrote a book called *Mathematical Foundations of Quantum Mechanics,* which was used in developing atomic energy.

In 1930, von Neumann came to the United States to take a position as professor of mathematical physics at Princeton University. He became interested in the use of large-scale computers and built one of the first modern electronic brains, called MANIAC (Mathematical Analyzer, Numerical Integrator and Computer). As a U.S. government advisor during World War II, he was influential in designing nuclear weapons and missiles.

Von Neumann had many intellectual interests, but his greatest joy was solving problems. Sometimes he would become so wrapped up in a problem while traveling that he would have to telephone his wife to find out why he had taken the trip. Because of John von Neumann's ability to solve problems, our mathematical horizons have been extended.

New electronic computers and new ways of solving problems with computers have been designed by mathematicians. These computers have been used to design automatic control of factories. These computers have also been used to extend mathematical knowledge; for example, a computer has made a list of all prime numbers below 46,000,000. However, a computer cannot create new mathematical ideas, and cannot solve the simplest problem unless a mathematician first "tells" it how to solve the problem.

A new algebra that applies to motions, numbers, or spaces is called *group theory*. We have already mentioned a new geometry, topology, that is concerned with relationships of points and lines but is not concerned with shape or size. A new arithmetic has been created to compute with new numbers called *quaternions*. *Sets* are being used in a new way to solve logical problems with electric circuits.

How does a mathematician create new mathematical ideas? New mathematics often comes from plain curiosity about a problem or an idea. By intuition, estimation, guess, experimentation, recollection, visualization, the mathematician searches for a clue to the mystery. Experiments in mathematics often require no tools, no equipment, and usually no materials other than paper and pencil. From the results, the mathematician states a conclusion. Then he uses deduction to prove that his conclusion is correct.

One of the most stupendous surveys of all mathematics created to date is *The Elements of Mathematics*, which started out as a joke. It consists of the writings of a fictitious Frenchman, Nicolas Bourbaki. Bourbaki was created by the imagination of a group of French mathematicians led by André Weil, one of the greatest modern-day mathematicians. This group began writing technical notes to mathematical journals under the pen name of Bourbaki. In about 1940 "Bourbaki" began compiling an exposition of all mathematics. Whenever new mathematical ideas are discovered, they will be added to Bourbaki's *Elements*.

EXERCISE SET 14
More Mathematical Thinking

1. Write the squares of the whole numbers from 1 through 10. Beginning with 2, write all numbers that can be obtained as sums of two

or three or four of these squares. For example, $13 = 9 + 4$ and $33 = 16 + 16 + 1 = 25 + 4 + 4$. Can you find a number less than 100 that requires more than four squares as a sum?

2. A $3'' \times 3'' \times 3''$ cube painted red on all sides is cut into 27 cubes $1'' \times 1'' \times 1''$. How many of these cut cubes are painted on 3 sides? 2 sides? 1 side? no sides? How did you solve the problem—by experiment or by using your imagination?

New Uses for Old Mathematics

Although we have noted many practical uses for mathematics, mathematicians are not always concerned with applications when they produce new mathematics. When Gottfried Leibniz (1646-1716), the great German mathematician, created the binary numeration system, it was of no practical value. Now this system, in which we can write any number using only the figures 0 and 1, is the one that is used in the operation of electronic computers. When the existence of a number such as $\sqrt{-5}$ was suggested, it was called an *imaginary number* because it seemed such a ridiculous idea. After all, it was thought, there is no number which, when multiplied by itself, produces -5. But now imaginary numbers are needed to solve problems about electricity. When the German mathematician Bernhard Riemann (1826-1866) suggested that the shortest distance between two points is a curved line, it was considered a foolish statement. Now nuclear physicists are using this idea and other ideas of non-Euclidean geometry in their research.

Through the study of pattern, mathematicians have discovered ideas that later have been found very useful. For example, the Greeks studied the ellipse over a thousand years before the German astronomer and mathematician Johannes Kepler (1571-1630) used their ideas to predict the motion of the planets. Now, the formulas for predicting the motion of the planets are applied to artificial satellites and will be useful in acquiring knowledge about space travel.

The new mathematical ideas being created today may not have application for years or even centuries. But most mathematicians are confident that if they create good mathematics it will someday be found useful. The Polish-American mathematician, Samuel Eilenberg, illustrates this when he compares his work with that of a creative tailor. He says, "Sometimes I make coats with five sleeves, other times with seven sleeves. When it pleases me, I make a coat with two sleeves. And if it fits someone, I'm happy enough to have him wear it."

Some Unsolved Mathematical Problems

Although we usually think of mathematics as an "exact science" that solves all problems, there are a number of mathematical problems that are still mysteries to mathematicians.

The Prime Number Mysteries. Some of the oldest unsolved problems involve prime numbers. For example, no one has been able to write a formula or system that will test whether or not a given number is a prime number. There must be some way of forming prime numbers, but no one has yet been able to find a systematic way to do it.

Another mystery about prime numbers is asked by the question, "Is there an infinite number of prime pairs?" A prime pair is a pair of prime numbers whose difference is 2; for example, (3, 5), (11, 13), (41, 43). These prime pairs seem to occur throughout our number system. No one has been able to find how many there are or to discover a formula to locate them. But, on the other hand, no one has been able to prove that there is a number beyond which there are *no* prime pairs.

Goldbach's Conjecture. "Is every even number the sum of two primes?" This is still another mathematical mystery. In 1742, the German mathematician C. Goldbach wrote a letter to his friend, the great Swiss mathematician Leonhard Euler (1707-1783), in which he made the conjecture that every even number except 2 was the sum of two primes. This was an interesting statement that was true for every even number he examined, but he could not prove that it was a true statement for all even numbers.

If you try some even numbers you will find that it always works; for example, $4 = 2+2$, $6 = 3+3$, $8 = 3+5$. *No* even number has been found that is not the sum of two primes. But this is no proof that *every* even number is the sum of two primes. If you could find one even number that is not the sum of two prime numbers, then the problem would be solved. Since no logical proof has been found for this seemingly simple problem, it is still one of the mysteries of mathematics.

Fermat's Last Theorem. Another famous mathematical mystery is called *Fermat's Last Theorem*. In the margin of a mathematics book, Pierre de Fermat (1601-1665), a famous French mathematician, wrote, "If n is a number greater than 2, there are no whole numbers a, b, c such that $a^n + b^n = c^n$. I have found a truly wonderful proof which this margin is too small to contain."

It was known long before Fermat's day that when $n = 2$, it is easy to find whole numbers x, y, z, such that $x^2 + y^2 = z^2$; for example, (3, 4, 5) or (5, 12, 13). (Such numbers, by the way, are called *Pythagorean numbers*.) But no one has ever been able to find positive integers x, y, z, such that $x^3 + y^3 = z^3$ or such that $x^4 + y^4 = z^4$. What Fermat was saying was that he could *prove* that no such numbers could be found.

After Fermat's death, his marginal note was discovered and mathematicians set out to find a proof for Fermat's theorem. No one has been successful in proving it. But, on the other hand, no one has been able to disprove it, either. This is strange because all of the other statements that Fermat claimed he could prove have been proved by later mathematicians. Actually, Fermat's theorem is of little real importance, but, as a result of attempts to prove it, some important ideas in modern mathematics have been developed.

The Odd Perfect Number Mystery. The ancient Greeks considered some numbers to be *perfect*. Perfect numbers are numbers which are equal to the sum of their divisors. The number 6 is such a number because $6 = 1 + 2 + 3$. Another perfect number is 28, since $28 = 1 + 2 + 4 + 7 + 14$. The next perfect number after 28 is 496. Others have been found and all of them are even numbers. No one has ever found an odd perfect number. But no one has been able to prove that every perfect number must be even.

Three Construction Problems. Some of the first unsolved problems in mathematics were these three famous constructions proposed by the Greeks, to be solved using only a pair of compasses and a straight edge:

1. Can you construct a circle with the same area as a square?

2. Can you construct a cube exactly twice the volume of a given cube?

3. Can you divide an angle into exactly three equal angles?

Mathematicians worked on these problems for many years before they found the solutions. However, the solutions were not what you might expect. The solution for each of these problems is the same; namely, that it is impossible to perform these constructions using only a pair of compasses and a straight edge.

How to Pack Spheres. A geometry problem that is still unsolved involves the packing of spheres such as ping-pong balls. How should spheres be packed in a box so that they use the least possible space? This is similar to a problem of drawing circles. How should circles be drawn or round objects like pennies be packed to cover the least surface? The arrangement for the circles on a surface has been found to be the following pattern:

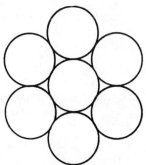

Figure 20

In packing spheres, this is also the best arrangement for the first layer. But nobody has solved the problem of how to arrange the second layer of spheres.

The Four-Color Map Problem. There are also unsolved problems in the field of topology. One is the four-color map problem. How many different colors are needed to make a map so that countries with a common border are shaded differently? The drawing below illustrates some possible maps.

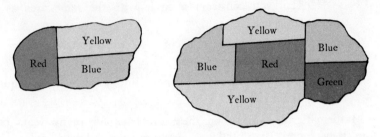

Figure 21

This is a real mystery to map makers and to mathematicians. They have not been able to draw a map that needs more than four colors. But at the same time they have not been able to prove that four colors are *enough* for any possible map.

Every day mathematicians and scientists are working on problems that seem to be unsolvable. The answer to some of these may be that the solution is impossible. To others the answers will be new ideas that will open new worlds of mathematics. Maybe you will be the one to become famous by finding the solution to one of these problems.

EXERCISE SET 15
Experimenting with Some Mathematical Mysteries

1. Write all the prime numbers from 1 to 50. Find the differences between the primes. Can you find a formula that will give you any prime?

2. a. What prime pairs can you find less than 100?
  b. How far apart are these pairs?
  c. Are these prime pairs equally spaced?

3. What two prime numbers have each of these even numbers for their sum?
  a. 8   b. 26   c. 18   d. 48

4. Do these sets of numbers satisfy the equation $x^2 + y^2 = z^2$?
  a. $x = 3, y = 4, z = 5$ c. $x = 8, y = 6, z = 10$
  b. $x = 7, y = 24, z = 25$

5. Do these sets of numbers satisfy the equation $x^3 + y^3 = z^3$?
  a. $x = 1, y = 2, z = 3$
  b. $x = 2, y = 3, z = 4$

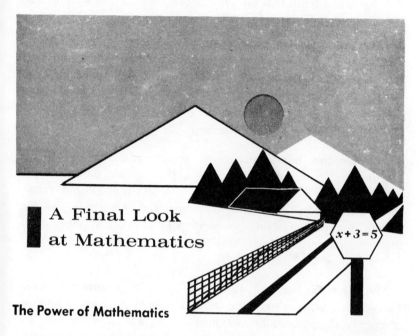

A Final Look at Mathematics

The Power of Mathematics

In this brief journey into mathematics we have tried to throw a spotlight on some of the interesting scenery that lies along the road. When you travel the road more slowly in the work that you will do in mathematics classes, you will have an opportunity to look more carefully at the trees, the plains, the structures that line that road. But even in this quick trip, we hope that you have been able to see that these are some of the remarkable qualities that mathematics possesses:

Mathematics provides an ideal model for logical reasoning. The deductive logic of mathematical proofs is considered the best method known for proving that a new idea is true. When the inductive method is used to establish facts, the conclusion is usually stated in terms of the mathematical theory of probability.

Mathematics has clarity and conciseness of expression. For example, the expression $\{ x \mid x + 3 = 5 \}$ is a short way of saying, "Find all values of x which make the sentence $x + 3 = 5$ a true statement."

Mathematics has certainty of conclusions. For instance, $2 + 3 = 5$ is a truth that always has been true and always will be true as long as 2, 3, and 5 are counting numbers.

Mathematics has novelty and a variety of fields. Mathematics may be concerned with such different things as maps, motion, games, music, art, probability, infinity, philosophy, and numbers. Who would ever expect that a mathematician would be interested in paper folding? But the folding of paper flexagons has intrigued many mathematicians.

Mathematics has abstractness which adds to its power. Geometry discusses points and lines, but no one has ever seen a point or a line. No one can make an object that has more than three dimensions, but mathematicians use expressions for the fourth dimension. No one can write all the prime numbers, but mathematicians can prove that there are an infinite number of prime numbers.

Mathematics has the power to predict events. In 1905 Einstein was able to write the formula that predicted the amount of energy that would be obtained by an atomic explosion. In astronomy it is possible to predict an eclipse of the sun using the formulas that give the motion and position of the heavenly bodies.

Mathematics can measure amounts indirectly. Mathematicians have measured the distance to the sun and the temperature at the middle of the sun without getting closer than about 92,000,000 miles. Even in 230 B.C., Eratosthenes measured indirectly the distance around the earth before it was even known that the earth was round!

Mathematics has unlimited opportunity for creativeness. Just as no one has composed the most elegant poem or painted the most beautiful picture, so no one has invented the ultimate mathematical structure. Every field of mathematics, from arithmetic to topology, gives ample opportunity for the creation of new ideas.

Mathematics has more permanence than any other field of knowledge. It is the only science in which the major theories of 20 centuries ago are still true and useful. The value of π has always been 3.14159 . . . and always will be (even though a state legislature once tried to change it to 3!). The musical scale that the ancient Greek philosopher and mathematician Pythagoras established, in which vibrations producing tones have the ratios $\frac{1}{2}$, $\frac{2}{3}$, and $\frac{3}{4}$, continues to be a basic scale in music.

Mathematical curves and surfaces have a balance and symmetry that are as pleasing as a masterpiece in art. From the simple circle to the complex hyperbolic paraboloid, from the golden section to pendulum patterns, mathematical curves are basic in art and architecture, in advertising, and in the graphic arts.

Mathematics is found in the designs and laws of nature. From the spiral of a snail's shell to the symmetry of a snowflake, from the bee's hexagonal cell to the elliptical orbits of the planets, mathematical curves and geometrical designs occur in the world of nature. Similarly, the distance of fall of a raindrop ($S = \frac{1}{2}gt^2$) and the energy in an atom ($E = mc^2$) can be expressed by mathematical formulas.

For Further Reading

This part of the book can do no more than touch upon the many interesting ideas in mathematics. We hope that you will want to go ahead on your own in mathematics and read more about it. The books in this annotated list will give you more information about many aspects of mathematics.

Basic Ideas in Mathematics

ADLER, IRVING, *The New Mathematics*. The John Day Co., 1958. This is a brief survey of new and old mathematics from a modern point of view, written in simple language.

ALLENDOERFER, C. B., AND OAKLEY, C. O., *Principles of Mathematics*. McGraw-Hill, 1955. This is an introductory college text which assumes little background in mathematics. It begins with work on logic and the number system, treats groups and Boolean algebra, and contains an introduction to analytic geometry, calculus, statistics, and probability.

BANKS, J. HOUSTON, *Elements of Mathematics*. Allyn and Bacon, 1956. The book is centered around four broad aspects of elementary mathematics: number, proof, measurement, and function. It is designed to make more meaningful the fundamental concepts and techniques of mathematics.

BOEHM, GEORGE A. W., *The New World of Math*. Dial Press, 1959. This book provides an exciting look at new creations and new applications in mathematics.

CUNDY, H. M., and ROLLETT, A. P., *Mathematical Models*. Oxford University Press, 1952. This is a source book for material on student projects. It covers a variety of topics, such as paper folding, curve stitching, polyhedra, models, linkages, and machines for solving equations.

KEMENY, J. G., SNELL, J. L., and THOMPSON, G. L., *Introduction to Finite Mathematics*. Prentice-Hall, 1957. This is an introduction to set theory, probability, logic, matrix theory, game theory, and linear programming.

The History of Mathematics

BELL, E. T., *Men of Mathematics*. Simon and Schuster, 1937. This voluminous book is a series of stories about the mathematicians who have made major contributions to mathematics.

HOGBEN, LANCELOT, *Wonderful World of Mathematics*. Random House, 1955. This is a beautifully illustrated history of mathematics written for young people. The color illustrations alone will often give more information than whole paragraphs in other books.

KLINE, MORRIS, *Mathematics in Western Culture*. Oxford University Press, 1953. The object of this book is to advance the idea that mathematics has been a major cultural force in the development of our western civilization. It shows how mathematics is the foundation for art, music, science, and reasoning.

REID, CONSTANCE, *From Zero to Infinity*. T. Y. Crowell Co., 1955. This book tells the history and extraordinary things about each of the digits 1 to 9, zero, and infinity. It is a fascinating story about our natural numbers.

Essays and Stories

ABBOTT, EDWIN A., *Flatland*. Dover Publications, 1952. This is a science-fiction classic about life in a two-dimensional world; it has political, moral, satiric, and humorous overtones. No previous knowledge of mathematics is required.

FADIMAN, CLIFTON, *Fantasia Mathematica*. Simon and Schuster, 1958. This is an anthology of essays and poems which involve mathematical ideas, especially topology and the fourth dimension.

NEWMAN, JAMES R., *The World of Mathematics*. Simon and Schuster, 1956. This is a four-volume anthology of famous mathematical essays from Archimedes to Einstein. The selections range from the humorous to the sublime. Introductory comments add to the appreciation of the articles, which show the tremendous range and contribution of mathematical fields.

OGILVY, C. STANLEY, *Through the Mathescope*. Oxford University Press, 1956. For the student who has no advanced mathematical background or special aptitude, this is a fascinating introduction to mathematics. It includes dramatic sidelights of mathematics such as probability, topology, unusual curves, and oddities.

Ravielli, Anthony, *An Adventure in Geometry*. Viking Press, 1957. In concise paragraphs and attractive graphic art, this book describes geometric shapes and their occurrence in nature. It includes sections on non-Euclidean geometry, projective geometry, and symmetry in simple yet interesting language.

Mathematics for Recreation

Bakst, Aaron, *Mathematics: Its Magic and Mastery*. D. Van Nostrand Co., 1952. An inexhaustible source of amusement in mathematics and an understandable discussion of algebra, geometry, trigonometry, calculus, etc. It includes sidelights on a variety of mathematical topics such as codes, chance, topology, *e*, number bases.

Heath, Royal, *Mathemagic*. Dover Publications, 1953. One of the original recreational mathematics books. Includes some of the most unusual puzzles and tricks with numbers. Besides the "magic," the book contains unusually interesting treatments of numbers called "symphonies" and short cuts called "Arithmeticklish."

Kasner, Edward, and Newman, James, *Mathematics and the Imagination*. Simon and Schuster, 1940. This classic treats such topics as geometric fallacies, paradoxes, topology, probability, and non-Euclidean geometry, in an informal, amusing style.

Northrup, Eugene, *Riddles in Mathematics*. D. Van Nostrand Co., 1944. A book of mathematical riddles, unusual problems, and paradoxes in arithmetic, algebra, geometry, logic, trigonometry, probability, and higher mathematics.

PART II

The World of
Measurement

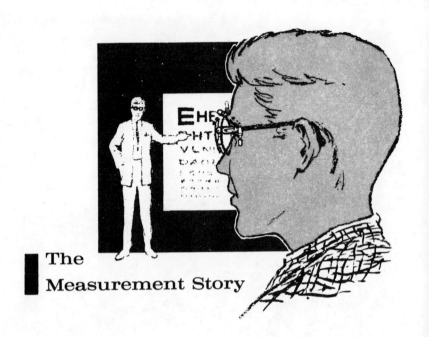

The Measurement Story

What Is Measurement?

How much air did you breathe today? How many calories are there in a candy bar? How far is it to the moon? How much light is needed for you to read this page? All these questions can be answered by measurements. The food we eat, the water we drink, and the air we breathe have been measured. Light, heat, electricity, and sound have been measured. Perhaps your eyesight, your intelligence, or your running speed also has been measured.

But what is meant by measurement? How are things measured? Who invented the units used to measure things? How do we compute in measured quantities? These are questions you need to know if you are going to be up-to-date in our mathematical world.

Suppose you go to a grocery store and buy a bag of 17 apples having a total weight of 6 pounds and a total cost of 90 cents. Three measurements are given in this situation. In each measurement, a *number* was used to describe some property. The *quantity* of apples was described by *17*; the *weight* in pounds was described by *6*; and the *cost* in cents was described by *90*. Also, associated with each number is a *unit of measure:* 17 *apples*, 6 *pounds*, and 90 *cents. A measurement associates a number with some unit of measure in order to describe some property of a person or thing.*

When we talk about measurements, we must keep things and their properties separate. *Things* are natural objects like boxes, footballs, or human beings. *Properties of things* refer to characteristics such as color, length, weight, or intelligence. We always measure the properties, not the things. We do not measure John; we measure John's height, weight, or temperature. In this way, measurement gives us information. To describe John as being 6 feet tall gives more information than to say John is a tall boy. To say that Nelly is overweight is not as clear as to say Nelly weighs 210 pounds.

An important part of measurement is the unit of measure. We use a length like this _____ , called an inch, to measure the length of a line. We say the length of line AB, below, is 3 inches.

A _____ B

Line AB can be divided into 3 parts, each 1 inch long. The measurement of line AB is, then, 3 inches. When we measure a quantity, we find how many of the appropriate units of measure it contains.

The Birth and Growth of Measurement

The history of man is, in part, a history of measurement. Measurement started by comparing things. By comparison, man was able to determine that one herd of animals was larger than another. When numbers were developed, this comparison was made by counting. Later, man developed better units of measure.

The first measurements made by using units of measure were very simple. A man used his foot, hand, thumb, or step to measure length. In this way, he always had his measuring scale handy. He used stones or kernels of grain to measure weight. He used the sun or the moon to tell the time or the season.

Now, these units of measurements were not always alike. One man's foot was larger than another's; some stones were heavier than others; and some days were longer than others. So, when man began to build homes, travel in ships, trade products, and divide the land, these natural ways of measuring weren't good enough. Now man needed standard scales for measures.

Whenever measurements were improved, new discoveries were made and better products resulted. This has been true as our civilization has grown. In the eighteenth century, the men in the factories making cannons and cannon balls found that they had to be very precise in their work. The cannon balls had to be made to fit the cannon barrels. If the balls were too small, the force of the powder explosion would leak past the ball and the shot would never reach the target. Of course, the cannon ball couldn't be larger than the barrel, either. Both the barrel and the cannon ball had to be made to accurate measurements. For the same reasons, the English inventor James Watt had to have a piston fit a cylinder exactly before the first steam engine would work. Man needed devices that could help him make accurate measurements.

Today, modern industry must make thousands of parts that meet precise measurements. Accurate measuring instruments make these precise measurements possible. Thus, it is possible to make cars, airplanes and machines that work for years and years. Some of the new measuring instruments are quite fascinating. For example, a device called a *sonigage* will measure the thickness of a surface from one side, whether it is the shell of an egg or the steel hull of a battleship.

Which contains more ground, a rectangular plot 18 feet by 22 feet, or one 30 feet by 10 feet? You can solve this problem easily by doing some simple computation. But it was a long time before man developed the ideas needed to answer such a question. As man learned more about computing with measurements, he was able to create new ideas and improve his way of life.

Convenient units of measure, standard measuring scales, precision measuring instruments, and a system for computing in measured quantities are essential parts of a useful system of measures. This means that measurement is a combination of mathematics, science, and mechanics. As mathematics, science, and mechanics have improved, measurements have improved. And with better measurements men have been able to create a world of new ideas, new materials, and new machines.

How We Measure

Measurements are used to determine how much or how little, how great or how small, how much more than or how much less than. Such determinations are often made by using the numbers which we read from instruments like a ruler, thermometer, clock, or measuring cup. These numbers represent the number of units of measure contained in a certain quantity. Measurements made in the same unit can be compared to see which measurement is the larger. Whenever we state or record a measurement, we should state the unit being used. For example, if someone says his little brother is 10, we might wonder if he means 10 days, 10 weeks, 10 months, or 10 years.

When we make a measurement, we usually make it in one of three ways — by counting, by direct use of a measuring instrument, or by indirect computation. Counting is often used to determine a measure when the unit of measure cannot be expressed in fractional parts. We count the boys at a party and we score points in a basketball game. In each of these situations, the measure is determined by counting, and a fraction would have no meaning. For example, it is meaningless to speak of counting $\frac{1}{2}$ boy or scoring $\frac{3}{4}$ of a point in a basketball game. In each of these counting measures, the unit of measure is one object, such as one boy or one point. Counted measures are usually *exact*. For example, the statement "The baseball team scored seven runs," would mean that the team scored exactly seven runs — not six runs or eight runs. However, sometimes counted measures are *inexact*. It is unlikely that you would give an exact number for the blackbirds in a flock or the population of a city. Instead, you would probably give an approximate value obtained by guessing, estimating, or rounding off an exact numerical result.

When a measurement is made by comparing an object to be measured with the units on a measuring instrument, we have direct measurement. When we use a ruler to measure the length of a sheet of paper, and a measuring cup to measure the amount of sugar in a box, we are using instruments to get *direct* measurements.

When we cannot apply a measuring instrument directly to the object to be measured, we use *indirect* measurement. We will have more to say later about indirect measurement and the methods by which it is obtained.

Counted measures for which fractions have no meaning are called *discrete data*. Values which are obtained by using a measuring instrument like a ruler are called *continuous data. Such measurements which give us continuous data are never exact.* Weight measures are examples of continuous data. You can measure your weight in any fractional unit you like, such as $105 \frac{5}{16}$ pounds. If you use a more precise scale, you may find your weight to be $105 \frac{9}{32}$ pounds.

A scale of different precision would probably give a different result. No matter how precise a measuring scale you have, it is always *possible* (but perhaps not practical) to devise one that would be more precise. Later in this booklet, we will say more about the inaccuracies involved in continuous data.

EXERCISE SET 1
Describing Measures

1. Would measurements of the following quantities give rise to discrete or continuous data?
   a. The number of people at a football game
   b. The heights of the boys in your school
   c. The price of a hot dog
   d. The number of baseballs used in a major league game
   e. The amount of water used in your school
   f. The time spent in school

2. Would measurements of the following quantities be exact or approximate?
   a. The distance from Chicago, Illinois to St. Louis, Missouri
   b. The weight of an apple
   c. The average number of apples that can be placed in a bushel basket
   d. The amount of milk in a paper cup
   e. The money in your bank account
   f. The number of members of the Democratic and Republican parties

3. Through the years, a variety of special units have been used to express counting measurements. What number is indicated by each of the following expressions?
   a. Span of horses
   b. Quire of paper
   c. Score of years
   d. Quartet of singers
   e. Brace of ducks
   f. Gross of boxes
   g. Ream of paper

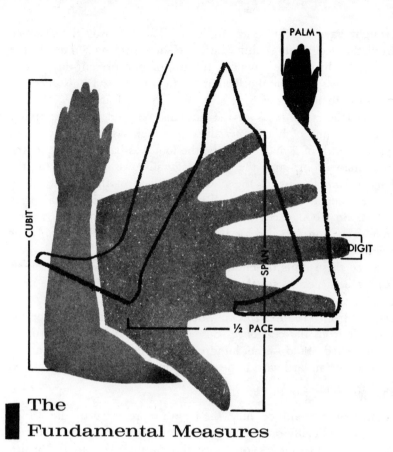

The Fundamental Measures

The Story of Units of Length

Length measurements, such as the distance between a cave and a stream, were among the first measurements man needed. When he began to build homes, to farm, to trade, or to travel, measures of length became more important.

The early measures of length were based on parts of the human body. The breadth (width) of a finger was called a *digit* and the width of a human hand was called a *hand* or a *palm*. The distance from the end of the thumb to the end of the little finger when the hand was stretched out fully was called a *span*. The distance from the elbow to the end of the longest finger when the hand was held straight gave a unit called the *cubit*. The *pace* was the distance from the heel to the point where the same heel next touched the ground. This was equivalent to two of our present-day paces. Thus, it was easy for the Roman soldiers to measure distances by counting their paces as they marched through Europe.

The lengths of parts of the human body led to some of our standard units of linear measure. Our inch was probably based on the width of a man's thumb; the foot was based on the length of a man's foot; and the yard originally was described as the length from the tip of the nose to the end of the thumb of an outstretched arm. In ancient Rome, the pace was equivalent to 5 feet. So 1,000 paces was 5,000 feet or almost as much as our mile. The Roman word for 1000 is "mille," which is probably the origin of our word "mile."

Since these units of length varied from one person to another, it became necessary for the ruler of a country to describe the units more accurately. For example, King Henry I of England, who reigned from 1100 to 1135 A.D., decreed that a yard should be the distance from the end of *his* nose to the end of the thumb of *his* outstretched hand. One hundred years later, King Edward I ordered a standard yard to be made from an iron bar and declared the foot to be $\frac{1}{3}$ of this length.

In the nineteenth century, the length represented by the British yard was officially described as a certain distance, at 62° Fahrenheit, between two marks on a special bronze bar kept by the British government. The United States borrowed the yard as a basic unit of length measurement from England. However, the present United States yard is described in terms of the metric system of measure, and therefore does not have exactly the same standard as the British yard.

EXERCISE SET 2
Working with Units of Length

1. Answer the questions that pertain to each of the following examples of measurements taken from famous literature.
   a. In Tennyson's poem "The Charge of the Light Brigade," we read "Half a league, half a league, half a league onward." What is the length of half a league in miles?

b. In Shakespeare's "The Tempest," there is a line: "Full fathoms five thy father lies." How deep is a fathom in feet?

c. The fifteenth verse of the sixth chapter of Genesis gives the dimensions of Noah's ark. It reads, "The length of the ark shall be 300 cubits, the breadth of it 50 cubits, and the height of it 30 cubits." What were the approximate dimensions of the ark in feet?

d. The Bible describes the dimensions of Solomon's temple. Find out what they were.

2. Obtain, from several people, measurements for the parts of the body described in the table below. Copy and complete the table, and compare the measurements made to the standard units to which they are related.

Description of measurement	First person	Second person	Third person	Fourth person
a. Standard unit — **Inch** width of thumb				
width of two fingers				
distance from knuckle to joint of thumb				
distance from joint to end of thumb				
b. Standard unit — **Foot** length of foot				
c. Standard unit — **Yard** distance from nose to thumb of outstretched arm				

3. A cubit is the length of a forearm from the elbow to the tip of the middle finger. Find the number of inches in a cubit measured from several people.

4. A pace is the distance between a point on one foot to the same point on the same foot when it next touches ground after a full step has been taken. Walk twenty paces, measure the distance in feet, and then divide by 20 to get the length of your pace.

The Story of Weight

The first measurement of weight was probably a comparison of two objects made by holding one in each hand and guessing which of the two was heavier. Sometime later, a wise man invented a simple weighing machine. This was a stick hanging by a rope tied around its middle. The objects to be compared were hung at opposite ends of the stick. If the objects weighed the same, the stick stayed level. If one object was heavier than the other, the stick tipped down on that side. Now we accurately weigh objects in the scientific laboratory by this same balancing system. The modern balance scale uses known weights on one side to balance the object being weighed.

The balance and weights system of weighing was probably started by the ancient Egyptians who used stones for weights. Even today, there is a unit of weight called the *stone*. A stone is equivalent to 14 pounds. The Babylonians used seeds of grain as weights, and several of the weighing systems in present use contain a unit called a *grain*.

Many of the units in the weight systems used in the United States and England developed from Roman measuring units. Today, we usually express the weight of an object in the *avoirdupois* system. The units in this system are pounds, ounces, drams, grains, and tons. A pound contains 16 ounces, an ounce contains 16 drams or $437\frac{1}{2}$ grains, and a ton contains 2,000 pounds. The avoirdupois system is not the only one used in the United States. The *apothecaries'* weight system is used by chemists and druggists to express the weight of drugs and medicines. In this system, a pound contains 12 ounces or 5,760 grains, an ounce contains 8 drams or 480 grains, a dram contains 3 scruples, and a scruple contains 20 grains. The *troy* system of weights is used in weighing gold, silver, platinum, coins, and certain jewels. The troy pound also contains 12 ounces or 5,760 grains, but an ounce is made up of 20 pennyweights, and a pennyweight is made up of 24 grains. The grain is the same weight in all three systems.

EXERCISE SET 3
Questions of Weight

1. Suppose you have eight coins of the same kind and same appearance but one of them is a counterfeit and is lighter than the others. How can you use a balance scale and find which coin is counterfeit in only two weighings?

2. Only five weights are needed for one side of a balance scale to weigh in grams from 1 to 31 grams. What weights should be used?

The Story of Units of Area and Volume

The amount of floor covered by a rug or the amount of land in a farm is measured in terms of area. In mathematics, we say that area describes the size of the interior of geometric figures.

How can we describe the size of the interior of Figure 1a?

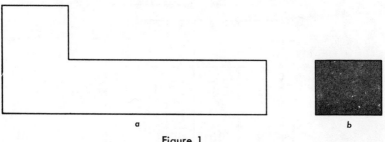

a   b

Figure 1

One way would be to compare it with the rectangle shown in Figure 1b. How many of these small rectangles will fit into the interior of Figure 1a? If we were to cut out several rectangles, the size of Figure 1b, and if we were very careful not to let the rec-

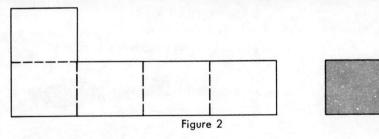

Figure 2

tangles overlap or extend outside the area enclosed by Figure 1*a*, we could cover the interior of Figure 1*a* as shown in Figure 2. Then we would say the interior or area of Figure 1*a* is equal to five unit rectangles the size of Figure 1*b*.

We could use any shape and size figure we wish as a unit of area measurement. After we have selected a unit of area, we measure the area of a figure by finding how many of the unit figures are needed to cover the interior of the figure being measured.

It has been found that the most convenient figure to use as a unit to measure area is a square. A square is selected that has a side which is a common unit of length — e.g., a square inch. The square foot, square yard, square meter, and square mile are other squares used as units of area measure. An acre is a unit of land area which is about as large as a football field. One acre equals 43,560 square feet.

Using square units makes it easy to compute the area of rectangles and other geometric figures if the dimensions are known.

For example, if the dimensions of the rectangle in Figure 3 are 4 inches by 3 inches, we say the area is 12 square inches, for the rectangle contains 12 squares having one-inch sides.

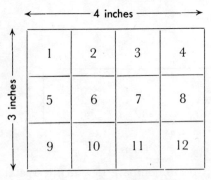

Figure 3

What is a simple formula for computing the area of a rectangle?

Some ancient measures of volume were barrels, sacks, bags or bunches. As our way of living has changed, we have changed the containers we use and the quantities we buy so these units have been discarded. We no longer buy a barrel of flour, a cran of herring, a hogshead of oil, a firkin of butter, or a hand of bananas.

Today we commonly use a cube with edges of a certain length as a unit of volume. If this length is one inch, it is called a cubic inch. For larger spaces, we use cubic feet or cubic yards and for smaller spaces we use cubic centimeters or cubic millimeters. If the dimensions of the block shown in Figure 4 are 4 inches by 3 inches by 2 inches, then the volume of the block is 24 cubic inches, for it contains 24 one-inch cubes.

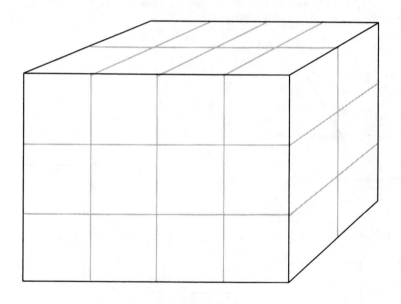

Figure 4

From this figure, you can easily express a formula for computing the volume of a box when you know its length (L), its width (W), and its height (H).

You are familiar with such special measures of volume for liquids as *gallons* of gasoline, *quarts* of milk, *pints* of cream, and *fluid ounces* of soda. Although you know that there are two pints in a quart and four quarts in a gallon, you perhaps do not know that a gallon in the United States is a volume of 231 cubic inches, or that a fluid ounce is $\frac{1}{128}$ of a gallon.

We also have special units, such as pints, quarts, pecks, and bushels, for measuring dry substances such as grains and fruits. However, a dry pint does not contain the same number of cubic inches as a liquid pint.

EXERCISE SET 4
Area and Volume Units

1. Find the formula for computing the area of each of these figures.
    a. triangle   b. circle   c. parallelogram

2. Find the area of each of the following figures in terms of the special unit of area given with the figure.

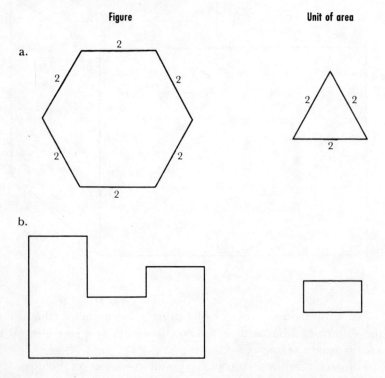

Figure   Unit of area

a.

b.

3. Express each of the following in a more common unit of measure.
    a. barrel   b. hogshead   c. firkin

4. If an automobile gasoline tank holds 20 U. S. gallons of gasoline, what is the volume in cubic inches?

5. Find out how much larger a dry pint is than a liquid pint.

6. The United States gallon and the British Imperial gallon are not equal in volume. See if you can find out which is larger.

7. How many pecks are there in a bushel? How many quarts are there in a peck?

8. The basic pattern for a cube is shown at the right. Use this pattern to build a cubic inch and a cubic foot. How many cubic foot cubes do you need to build a cubic yard?

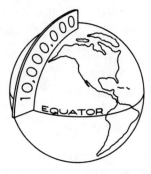

The Metric System

For over a hundred years, scientists have been using the metric system of measurements. It is much simpler and more convenient than our own system, but we are very slow to accept a new system. The United States and England are the only major nations that have not adopted the metric system. However, most scientists, government agencies, and many industries in both countries now use the metric system.

Many people believe the metric system was first suggested by French scientists in about 1790 as a basis for improving their system of linear measures. The standard unit of length, called a meter, was selected as the basic unit of length. This length was supposed to be $\frac{1}{10,000,000}$ part of the distance from the North Pole to the equator along a meridian. Now that we have better measuring instruments, we know this is not quite right, but the length of the first meter is still used as a standard. The meter is 39.37 inches long, or a little more than one yard.

The metric system, like our number system, is based on ten. The basic units of measure in the metric system are subdivided into tenths. The basic units are also multiplied by ten to form larger units. These units then are related like the place value of numerals in our numeration system.

The metric system is easy to remember because similar multiples or subdivisions of basic units use the same prefixes. Table 1 shows the meanings of the prefixes used in the metric system.

Table 1

Prefix	Significance	
mega	1,000,000	times the basic unit
kilo	1,000	times the basic unit
hecto	100	times the basic unit
deka	10	times the basic unit
deci	$\dfrac{1}{10}$	times the basic unit
centi	$\dfrac{1}{100}$	times the basic unit
milli	$\dfrac{1}{1,000}$	times the basic unit
micro	$\dfrac{1}{1,000,000}$	times the basic unit

You can see how the prefixes show the comparisons between metric units of length in Table 2.

Table 2

1 micrometer (or micron)	=	.000001	meter
1 milimeter	=	.001	meter
1 centimeter	=	.01	meter
1 decimeter	=	.1	meter
1 dekameter	=	10	meters
1 hectometer	=	100	meters
1 kilometer	=	1,000	meters
1 megameter	=	1,000,000	meters

To change from one unit to another, it is necessary only to either multiply or divide by a multiple of 10.

Units of area are squares with the basic units of length as the length of a side. The common unit is square centimeter or square meter. Scientists sometimes use exponents to show square units.

For example, 5 cm.^{2} means five square centimeters. The ^{2} is the exponent.

Units of volume are cubes with the basic unit of length as the length of an edge. The common unit is cubic centimeter or cubic meter. Using exponents, we can write seven cubic centimeters as 7 cm.^{3} Another abbreviation for cubic centimeter is cc. Volume is also expressed in terms of liters. Thus a milliliter (ml.) would be $\frac{1}{1000}$ of a liter.

The weight of one cubic centimeter of water at a temperature of 4° centigrade is the basic unit of weight of the metric system and is called a gram. Another unit is the *kilo*gram (1,000 grams), which equals 2.20 of our pounds. Likewise, we can work with milligrams ($\frac{1}{1,000}$ gram), and so on.

Most countries have a metal bar that is a copy of the standard that represents a length of one meter. However, a new standard for the meter has been proposed in terms of the wave length of the orange light given off by hot krypton. Krypton is a gas present in very small amounts in the air and is used in small amounts in neon signs. The new meter would be the same length as the old one but would now be described as being 1,650,763.73 times the wave length of krypton 86. This is about 100 times more precise than the old standard. The distance between two scratches on a metal rod is not very precise. The scratches themselves have breadth which could affect the standard length. Another advantage of the new standard is that it can be produced in any good laboratory.

The common relationships between the metric system and the United States system of measure are shown in Table 3.

Table 3

1 meter = 39.37 inches
1 inch = 2.54 centimeters
1 kilometer = 1000 meters = 0.62 mile or about $\frac{5}{8}$ mile
1 gram = 0.035 ounce (avoirdupois)
1 liter = 1.057 liquid quarts
1 kilogram = 2.20 pounds

We can use the information in Table 3 to change measurements from one system to another. For example, Table 4 shows the number of centimeters that correspond to various lengths expressed in inches.

Table 4

Number of inch units (I)	1	2	3	4	5
Number of centimeter units (C)	2.54	5.08	7.62	10.16	12.70

If I represents the inch units and C the centimeter units, what formula can you write to represent the relationship expressed by the table? Notice that inches are larger than centimeters. To change a measurement in inches to centimeters, we multiply the number of inches, the longer unit, by 2.54 to get the number of centimeters. To change a measurement in centimeters to one in inches, we divide the number of centimeters by 2.54.

In order to keep our measures standardized, the federal government has set up the National Bureau of Standards. This bureau in Washington has copies of the original standards of the meter and kilometer so that *precise* measuring instruments can be checked for accuracy. In checking instruments this bureau uses devices that can measure very precisely. For example, one instrument can measure how much a steel bar bends when a fly sits on it and another can detect the heat of a man's body a half mile away. This bureau checks devices like scales, gasoline pumps, milk cartons, and electric meters so that we are sure to get the right amount when we make a purchase. This bureau also makes tests of products like shoes, cloth, plastic, to see that the government buys good quality merchandise. This bureau and many other laboratories are building new and more accurate measuring instruments so that new ideas will be discovered and new inventions made. Our life now and in the future depends a great deal on new and better ways of measuring everything in our world. For example, space travel will depend on new measurements that haven't even been invented today.

EXERCISE SET 5
Working with the Metric System

Find the missing quantity.
1. 1 decimeter = ? centimeters.
2. 38 millimeters = ? centimeters.

3. 5 dekameters = ? meters.
4. 2 square decimeters = ? square centimeters.
5. 2 cubic meters = ? cubic centimeters.
6. 8 inches = ? centimeters.
7. 24 miles = ? kilometers.
8. 6 kilograms = ? pounds.
9. 4 liters = ? liquid quarts.
10. 10 pounds = ? kilograms.
11. 12 kilometers = ? miles.

The Story of Time Measurement

What is time? Although this may seem like an easy question, it is not, for it is difficult to accurately define time. We know that time can be used to help describe and locate certain events or situations, but this does not tell us exactly what time is. But since one can locate events in time, just as we can locate objects in space, Einstein called time the fourth dimension.

Time has always been important to man, and was one of the first things he measured. Early man measured time by the rising of the sun and the phases of the moon. Thus, he spoke of events happening a certain number of "suns" or "moons" ago. This practice probably led to the first calendars of the ancient Babylonians and Egyptians. The story of the development and progress of calendars is an important and interesting one.

Eventually, clocks were developed to measure shorter periods of time. One of the first clocks, the sundial, used the sun's shadow to show the time of day. But this was not useful for night time or even on cloudy days. The hourglass was more useful and is still used by some housewives to time the boiling of eggs. These glasses are made so that sand will flow from one end to the other in a given time. The Chinese water clock used water instead of sand. Water dripped from one container to another to measure the passing of time. The early Anglo-Saxons in England used a candle to measure time. As the candle burned, it melted marked segments which showed how long the candle had been burning. The invention of gears, the discovery of the laws of operation of the pendulum, and the use of electric current has made it possible to invent more accurate clocks.

All the present units of time are exactly defined in terms of the movement of the earth with reference to the stars. Therefore, the measurement of time is closely related to the science of astronomy.

EXERCISE SET 6
Measuring Time

1. Find all the information you can on the Julian calendar and the Gregorian calendar. Which calendar is used the most in the world today?

2. The Italian astronomer and physicist Galileo (1564-1642) learned the secrets of a pendulum. He found that the time it took a pendulum to complete a swing depended on the length of the pendulum. The longer the pendulum, the slower it swung. See if you can discover the length of a pendulum that will complete a swing in one second.

3. Make pendulums of different lengths out of string and weights such as fishline sinkers. Measure the time for a complete swing. Try lengths such as $2\frac{7}{16}$ inches, $9\frac{3}{4}$ inches, $21\frac{15}{16}$ inches, 39 inches, and 13 feet.

4. Measure the height of a bridge or building by timing a pebble dropped from the top. The formula to use is $s = 16t^2$ where s is the distance in feet and t is the time in seconds.

The Measurement of Angles

Angles are used in many ways. For example, angles are measured in navigating an airplane, in building a house, in mapping a new park, and in computing the distance to the moon.

The figures below are illustrations of angles.

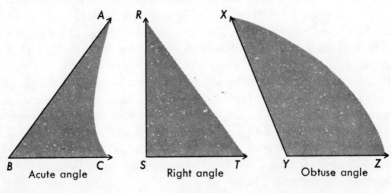

| B   Acute angle   C | S   Right angle   T | Y   Obtuse angle   Z |

Figure 5

Early in history, angles were probably visualized as figures formed by two radii of a circle, and this perhaps led to units of angle measure based on the circle. It is thought that the ancient Babylonians divided the circle into 360 equal parts called degrees because they believed a year to contain 360 days. The degree and its subdivisions (60 minutes in 1 degree, 60 seconds in 1 minute) are not the only units of measure for angles. Sometimes mathematicians or scientists use *radians* or *mils* to measure angles. These units are also based on the circle.

An angle of one degree is an angle which, when placed with its vertex at the center of a circle, cuts off $\frac{1}{360}$ of the circumference of the circle. An angle of one radian is an angle which, when placed with its vertex at the center of a circle, cuts off a length of the arc of the circle equal to the radius of the circle. An angle of one mil is $\frac{9}{160}$ of an angle of one degree or about 0.001 of a radian.

An angle is measured by finding the number of times it contains one of these unit angles.

EXERCISE SET 7
Working with Angle Measures

1. Calculate the number of degrees (to the nearest degree) in a one-radian angle.
2. How many degrees are there in angle x of the following figure?

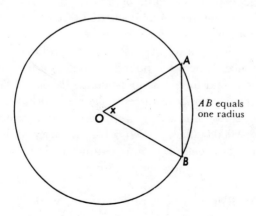

AB equals one radius

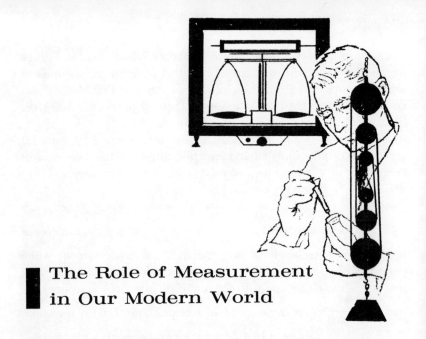

The Role of Measurement in Our Modern World

The Measurement of Force, Work, Energy, and Power

Measurements of length, weight, volume, and time are very important in science and industry as well as in everyday life. However, there are also many other kinds of measurements that are important in modern society. Measurements of force, work, energy, and power provide one example.

The units of *force* (a push or a pull) are the same as the units of weight (pounds, grams, and so on). This is natural since the weight of an object expresses the pull of gravity on the object. A very small unit of force, the *dyne*, is often used by scientists. A force of 980 dynes is equivalent to a force of one gram.

Work is done, in a scientific sense, when a force moves an object some distance. When one pound is lifted one foot, the amount of work done is said to be one foot-pound. If you lift 25 grams three centimeters, you do 75 gram-centimeters of work. Work is expressed in such units as *foot-pounds*, *gram-centimeters*, *ergs* (dyne-centimeters), or *joules* (10,000,000 ergs). *Energy* (the capacity to do work) is also expressed in foot-pounds, gram-centimeters, ergs, or joules. A foot-pound of energy is the energy needed to lift a one-pound weight one foot.

Power is defined as the rate of doing work. James Watt, the English inventor, found that an average horse could lift 550 pounds one foot high *in one second.* So an engine that can do 550 foot-pounds of work per second is called a *one-horsepower* engine. Since someone decided an average man can lift 90 pounds one foot in one second, this amount of energy is called one manpower.

The Measurement of Heat and Temperature

Many forms of energy that are important to us are measured with special units. Heat is one such form of energy. Heat is measured by the effect it produces. One unit of measure for heat is the *calorie.*

One calorie is the amount of heat needed to raise the temperature of one gram of water one degree centigrade. The amount of heat needed to raise the temperature of 1,000 grams of water one degree centigrade is called a large calorie. The large calorie is used to measure the heat energy in the food we eat. We use the energy from the food to do work, and since the work done in a certain amount of time can be measured in horsepower, it would be possible to determine the potential horsepower of a dish of ice cream.

Another unit of heat is the *British thermal unit (B.T.U.).* This is the amount of heat energy required to raise the temperature of one pound of water one degree Fahrenheit.

Most of the furnace and air-conditioning manufacturers use the B.T.U. to describe the performance of their products.

Notice that a measure of temperature is used to express the unit of measure of heat. But temperature and heat are not the same. Temperature measures the relative intensity of heat energy at a certain place. The intensity of the heat is expressed as a certain number of units (degrees) above or below a selected zero reference point. Two main temperature scales, Fahrenheit and centigrade, are familiar to us.

Measurements Related to Light

If someone told you to get a 200 foot-candle lamp, you might be confused as to what to do. Nevertheless, there is such a thing as a foot-candle. It is used by scientists to measure the brightness (intensity) of a light. One foot-candle is the intensity of light of a standard candle (a candle about one inch thick) at a distance of one foot from the flame.

Some of our units of length are related to light. One of the smallest units of length is the *angstrom*. It is used to measure the *wave length* of light or other similar radiations. The angstrom is a metric unit equal to $\dfrac{1}{10,000,000}$ of a millimeter or about $\dfrac{1}{250,000,000}$ of an inch. Astronomers use a measure called a *light year* (the distance light travels in a year, about 6 trillion miles). The longest unit of length is the *parsec* (3.26 light years).

The Measurement of Sound

Which is louder, the noise from a busy street corner or the noise from your school lunch room during noon hour? To answer this question, we must have some kind of measuring scale for the intensity of sounds. A scale using a unit called the *decibel* has been constructed for this purpose. A sound that can barely be heard by a normal ear has an intensity of 0 decibels. A sound that is so loud as to be painful to the normal ear has an intensity of about 120 decibels.

Musicians are quite concerned with the pitch of a sound. The pitch depends upon the frequency of vibration of the instrument producing the sound; the greater the frequency, the higher the pitch. Frequency is often measured in cycles per second.

The Measurement of Electricity

Some of the most fascinating and important measurements made in the world today are related to the field of electricity. Although we cannot fully discuss the units of measure used in electrical work, it is interesting to note some of the many types of measurements made in this field.

The *electron* and *coulomb* are used as units of electrical charge. The rate of flow of electrical charges is measured in *amperes*. The difference in the "pressure" of electrical charges between two points in a closed electrical circuit is measured in *volts*. The resistance to the flow of current is measured in *ohms*. Electrical power is measured in *watts* and electrical energy in *watt hours*. Because a watt is such a small unit, electric power companies generally charge for electrical energy by the *kilowatt hour*. The most commonly used unit of power is the *kilowatt*.

The Measurement of Clothing

How big is a size five shoe? What measurement is indicated by a size six hat? Clothing sizes are expressed in numbers that mean different things, depending upon the article of clothing. Gloves are measured by the number of inches around the palm of the hand and the fingers when the hand is made into a fist. Ladies' hat sizes represent the number of inches around the head. But a man's hat size gives the diameter of his head. Actually, it isn't quite this amount, because a man who blocked hats long ago couldn't figure very well and made a mistake. Ever since then, a man's hat has been $\frac{5}{32}$ inch smaller than the hat size.

Money Measurements

Isn't it remarkable that pieces of green paper containing some words, pictures and numbers can be traded for all sorts of goods? The story of the development of money systems is closely related to the story of measurement. When money was first made, the value of a coin was the value of the metal in the coin. The value, determined by the weight and purity of the coin, was written on the coin. Coins of large value were made of pure gold and coins of small value were tiny pieces of silver. At one time, the smallest coin was the "mite". This is the coin that is referred to in the Bible story of the widow's mite.

Now, our coins are worth more than the metal they are made of.

For example, the metal in a silver dollar is worth less than 50 cents. Of course, we now use paper money which is valuable because the government has printed on it a promise to pay its value to the bearer. The real value of money is what it will buy. When prices are high, money will not buy as much as it will when prices are low.

A country must have a system of money that can be used to facilitate trade. The government must select and define the basic money unit and then issue currency in various denominations of the basic unit. The conversion from the monetary measure of one country to that of another is often of great importance to a world traveler.

In this section and the several preceding it, we have illustrated a few of the important aspects of measurement in our present society. We could give many more examples of the use of measurements. For example, the weather forecaster measures air pressure, air temperature, and the amount of moisture in the air; businessmen get measures of buying trends; doctors make measurements that help describe the state of a person's health; and photographers must make measurements to get the proper camera setting for a picture. Measurement in the social sciences — for example, the measurement of intelligence — is another field which has not been discussed. Thus, we must conclude that knowledge of measurement is very important in our complex world.

EXERCISE SET 8
More About Modern Measurement

1. How much work is done in raising a three-pound block seven feet?

2. Find out your "manpower" or "horsepower" by measuring the time it takes you to move your weight up a certain height of stairs.

3. How many small calories are supplied by a substance that, when burned, raises the temperature of 12 grams of water 5°C.? (Assume that no heat is lost.)

4. Express 82 angstroms in millimeters.

5. Find the value of each of these foreign money units in United States money and name the country in which it is used.

    a. crown   b. pound   c. shilling   d. peso
    e. franc   f. lira   g. mark   h. ruble

6. The Bible mentions these coins: mite, shekel, talent. Try to find the value of each of these coins in United States money.

Working with Measurements

Estimating Measurements

We have seen that to make a measurement really means to associate a number with some unit of measure. Usually we get the number by counting or reading an instrument. Sometimes we get the number by computing. For example, when determining the volume of a room, we don't fill the room with water to see how many cubic feet it will hold. Instead, we find the dimensions of the room and do some computation. How we compute with the numbers we use in measurement depends on how they were obtained. We will talk more about this later.

To know what a measurement means, we need to know the nature of the unit of measure. Whenever we have worked with the unit of measure sufficiently, we usually have a good understanding of it. This is true of units such as a pound, candle power of a light, or decibels of sound. But units of energy, such as calories, or units of length, such as light years, are not so easy to visualize or imagine.

If you gain an understanding of the size of a unit of measure, you can use your knowledge to make estimates of the measures of things that often come up in day-to-day situations. It is often advantageous to be able to guess such measures as the distance of an oncoming car, the number of people ahead of you in a ticket line, or the weight of a piece of mail.

EXERCISE SET 9
Making Estimates

Make the estimate described in each of the following exercises, and then try to obtain factual information to check your estimate. If your amount of error is not greater than 10% of the actual value, you are doing well.

1. The time, in seconds, that a traffic sign stays red
2. The length of a block in yards, and the blocks in a mile
3. The acres of land in a block
4. The distance, in inches, a bicycle wheel travels in one revolution
5. The height, in feet, of a bridge
6. The height, in feet, of the tallest building in your town
7. The dimensions, in feet, of your classroom
8. The weight of a quart of milk in ounces
9. The gallons of water needed to fill your bathtub to a depth of ten inches
10. The diameter, in inches, of a human hair
11. The grains of sand in one teaspoon (If each person in your class counts a pinch of sand, the total count can be computed quickly.)
12. The length, in feet, of your family automobile
13. The flying speed, in miles per hour, of a bird
14. The cubic yards in your classroom
15. The weight of a cubic foot of gravel
16. The foot candles of light of the full moon
17. The temperature, in degrees Fahrenheit, of the water in a lake or stream near your home
18. The height of a cloud in feet
19. The weight of a shirt in ounces
20. The dimensions of an envelope in inches
21. The square feet of window space in your room
22. The weight of your mathematics book in ounces
23. The square yards of wallpaper needed for a wall of your room
24. The yards of cloth needed for making a dress
25. The pounds of roast beef for a dinner for six people
26. The weight of a dozen eggs in ounces
27. The number of pennies in a handful
28. The time required to walk $\frac{1}{2}$ mile
29. The cost of using your TV set for one hour

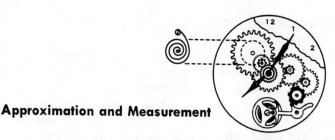

Approximation and Measurement

We know that a measurement is usually made by comparing the units represented on a measuring instrument with an object to be measured. This is the way we measure length with a ruler, volume with a measuring cup, or weight with a pan balance. Measurements obtained by direct comparison to standard units are called *direct measurements*.

We have no instruments that directly represent standard units of time. A clock contains a number of gears, wheels, and springs which produce a movement of hands around a dial. The distance moved by the hands (or the angle through which they move) indirectly represents the passage of certain units of time. The measurement of time with the clock is an example of *indirect measurement*. Another good example of indirect measurement is the measurement of temperature. When a fluid is heated it expands. If we put the fluid into a sealed glass tube, the amount it expands or contracts can tell us how much the temperature has changed. Thus, temperature is measured indirectly by measuring the level of a fluid in a tube.

Any time we make computations to obtain a measurement, we are measuring indirectly. For example, if we wished to find the area of this page of the book in square inches, we could do it directly by cutting squares one inch on a side and finding the number of squares needed to cover this page. However, it would be much simpler to measure the length and width of the page and then compute the area by multiplying those two values. The area measurement made in this fashion would be an indirect measurement.

Suppose Bill says that his brother Joey is 10 years old. Does this mean that Joey is exactly 10 years old? Very likely it doesn't. Most likely Bill means that Joey has passed his tenth birthday, but he probably does not mean that Joey is exactly 10 years old. The *number*, 10, which Bill reports is not approximate, but the *measurement* represented by 10 years is approximate. Thus, the number that appears in a measurement might not represent the exact value of the measurement.

Of course, whenever we measure with a measuring instrument, the results, as far as we know, are approximations to the actual value. Reading a scale or dial involves sighting, estimating, and judgment, and all three tend to give an approximate value to the measurement. Hence, no one ever has been or will be able to measure exactly the weight of an apple, the diameter of a hair, or the time to run a race. We can illustrate this point by measuring the line segment AB in Figure 6.

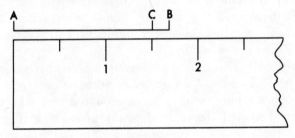

Figure 6

In measuring the length of AB with a ruler marked into $\frac{1}{2}$-inch lengths, we say line AB is $1\frac{1}{2}$ inches. As we can see, it is a little more than this, but we don't know how much more. In the same way, all measured quantities have errors involved. If we were to mark the scale in Figure 6 in units of $\frac{1}{4}$ of an inch, we could get a better measurement of the length of line AB. As we continue to increase the number of divisions on the ruler, we would expect the error in the measurement to decrease. But no matter how many divisions were marked on the ruler, we could not be certain that the end point of the line would exactly coincide with a mark on the scale. If a one-inch scale was divided into 16 parts, the end point of a line being measured by the scale might seem at first glance to line up exactly with some mark on the scale. But if each of the $\frac{1}{16}$-inch units was divided into two, four, eight, or more parts, the end of the segment might then seem to correspond to one of the new marks. No matter how many divisions you try to put on a scale, it is always possible to think of another smaller division that would give greater precision to a measurement. Thus, any measurement made with a measuring device is approximate. Exact numbers arise only from counting.

EXERCISE SET 10
Classifying Measurements

1. Which of the following are exact, or counted, measures and which are approximate measures?
    a. Baseball score
    b. Enrollment of your school
    c. Annual rainfall in Florida
    d. Time to run 100 yards
    e. Tickets sold for the concert
    f. Automobiles manufactured in June

2. Classify each of the following as examples of direct or indirect measurement.
    a. Obtaining the weight of an object with a spring scale
    b. Counting the number of nine-inch tiles in a floor to determine its area
    c. Finding the circumference of a circle by multiplying its diameter by 3.14
    d. Using the speedometer of an automobile
    e. Determining the passage of a year by the position of the sun
    f. Using a protractor to measure the number of degrees in an angle

Precision of Measurement

The measuring instrument we use to make a measurement depends on what we are measuring. For example, a thermometer is used to measure temperature, not air pressure. The measuring instrument we use might also depend on the size of the measurement we are making; we use the odometer of an automobile, not a ruler, to measure the distance between two towns. The scale or the dial on a measuring device is usually marked off to measure units or fractions of units. A ruler, for instance, might be marked off into eighths of an inch. The smaller the unit or fraction of a unit that a device can measure, the more precisely it will measure. *The precision of a measurement, then, depends on the size of the unit of measure used to make it.* The smaller the unit of measurement the more precise the measurement. A scale that is marked to give readings in ounces is more precise than the scale that is marked in half-pound units. For the purpose of precision of measurement, we would consider $5\frac{6}{8}$ inches more precise than $5\frac{3}{4}$ inches. The first measurement indicates the use of a ruler with marks for each eighth of an inch, the second one a ruler with marks for each fourth of an inch. A measurement such as $5\frac{6}{8}$ inches would not be reduced to $5\frac{3}{4}$ inches, because the denominator of the fraction of the recorded measurement would show us the size of the unit on the measuring scale. If the ruler being used to measure a length is marked off in eighths of an inch, the measurement should be read "to the nearest eighth of an inch." This means that the measuring unit is $\frac{1}{8}$ inch. If the length of a line is between $5\frac{6}{8}$ inches and $5\frac{7}{8}$ inches, it would be recorded as one or the other. Recording it as $5\frac{13}{16}$ inches would indicate more precision than is possible with a ruler marked off in eighths of an inch. Recording $5\frac{6}{8}$ inches as $5\frac{3}{4}$ inches would indicate less precision than is possible.

The *error of a measurement* which we make is the difference between the actual length and the recorded length. In the line AB in Figure 6, the segment CB beyond $1\frac{1}{2}$ inches represents the error made in the measurement of AB. If the scale had been marked in $\frac{1}{4}$-inch units, the error would have been less. Thus, the amount of error we make in a measurement depends, in part, on the scale of the measuring instrument used. An ordinary ruler might be used to measure length to the nearest $\frac{1}{8}$ inch. A micrometer might be used to measure the thickness of a hair to the nearest 0.001 inch. The possible error of a measurement made with the micrometer would be less than that of a measurement made by the ruler. We say that the micrometer has the smaller unit of measure (0.001 inch) and thus would give the smaller error of measurement. The error of a measurement should not be considered as a mistake, but should be recognized as a result of the limitations of the measuring device being used.

Use a ruler to measure the lengths of the lines in Figure 7.

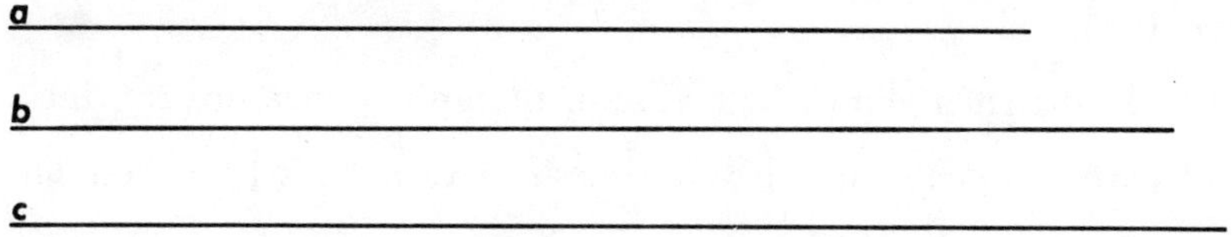

a _____

b _____

c _____

Figure 7

What is the length of each line in Figure 7 if measured to the nearest inch?

What is the length of each line in Figure 7 if measured to the nearest quarter inch?

You should have found all three lines to be 4 inches long when measured to the nearest inch, but you should have found the lengths to be $a = 3\frac{2}{4}$ inches, $b = 4\frac{0}{4}$ inches, and $c = 4\frac{1}{4}$ inches when measured to the nearest quarter inch. You cannot determine an exact error for each measurement, but it should be apparent that a line such as a, b, or c measured to the nearest inch would be recorded as 4 inches if it was between $3\frac{1}{2}$ and $4\frac{1}{2}$ inches in length. Thus, if we are measuring to the nearest inch, the greatest possible error is .5 inch or one half the smallest unit of measure on the measuring instrument. This greatest possible error is called the *absolute error*. This error always indicates the greatest difference that can be expected between the actual length and the recorded length. It does not mean that every measurement made by the device will have that amount of error. In measuring line b in Figure 7 the error was quite a bit less than one-half inch. The smaller the unit on the measuring scale, the smaller the absolute error and the more precise the measurement. Thus, writing the absolute error is a way to represent the *precision* of a measurement.

The absolute error is used to show the interval represented by a measurement numeral. Suppose the time to run a 100-yard dash was recorded as 9.8 seconds. This recording indicates that the time was measured to the nearest 0.1 second. The absolute error is one half the smallest unit of measure, $\frac{1}{2} \times (0.1)$ second or 0.05 second. The recorded time, 9.8 seconds, means that the running time was closer to 9.8 seconds than it was to 9.7 seconds or 9.9 seconds. This means that any time between 9.75 and 9.85 would be recorded as 9.8. How is this interval, 9.75 to 9.85, obtained from the recorded measure 9.8 and the absolute error of 0.05?

On the ruler shown in Figure 8, the unit of measure is $\frac{1}{4}$ inch. The recorded measurement for lines a, b, c, and d is $1\frac{1}{4}$ inches. The absolute error is $\frac{1}{2}$ of $\frac{1}{4}$ inch, or $\frac{1}{8}$ inch. The recorded measurement of $1\frac{1}{4}$ inches would apply to any measurement in the interval $1\frac{1}{8}$ inches to $1\frac{3}{8}$ inches.

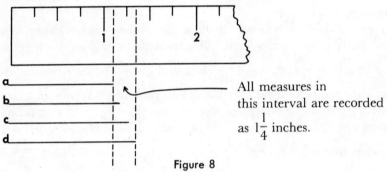

All measures in this interval are recorded as $1\frac{1}{4}$ inches.

Figure 8

The absolute error of a ruler marked off in eighths of an inch would be $\frac{1}{2}$ of $\frac{1}{8}$ inch or $\frac{1}{16}$ inch. The absolute error of a common micrometer would be $\frac{1}{2}$ of 0.001 inch or 0.0005 inch. This indicates how much more precise the micrometer is than a ruler marked off in eighths of an inch.

When a measured quantity is properly recorded, we can immediately determine the precision. For example, the measurements 2.5 inches and 158.7 inches are equally precise. The smallest unit of measure is 0.1 inch and the absolute error is $\frac{1}{2}$ of 0.1 inch or 0.05 inch. However, the measurements $135\frac{3}{4}$ inches and $7\frac{12}{16}$ inches are not equally precise. Which one has the smallest unit of measure? The smallest absolute error?

The precision of such measurements should be shown. Suppose you measured a line to the nearest $\frac{1}{4}$ inch and found the measurement closer to 5 inches than to $4\frac{3}{4}$ inches or $5\frac{1}{4}$ inches. The correct way to record this measurement is $5\frac{0}{4}$ inches, not 5 inches, for the reading $5\frac{0}{4}$ inches indicates a measurement to the nearest $\frac{1}{4}$ of an inch, while 5 inches indicates a measurement to only the nearest inch.

In a machine shop the greatest error allowed in the dimensions of a part is called *tolerance* and is specified in the blueprints. If a part that is to be $3\frac{5}{8}$ inches long may have an error of as much as $\frac{1}{16}$ inch, the tolerance is written $(3\frac{5}{8} \pm \frac{1}{16})$ inches, read "$3\frac{5}{8}$ inches plus or minus $\frac{1}{16}$ inch." This means that the part can be used even if it is as long as $(3\frac{5}{8} + \frac{1}{16})$ inches or $3\frac{11}{16}$ inches or as short as $(3\frac{5}{8} - \frac{1}{16})$ inches or $3\frac{9}{16}$ inches. The range $3\frac{9}{16}$ inches to $3\frac{11}{16}$ inches is called the *tolerance interval*. By specifying the tolerance allowed in making parts in factories, it is possible to have mass production of products such as automobiles and cameras. The parts can be made in different factories but will all fit at assembly because they meet the specified measurements.

EXERCISE SET 11
Finding Possible Errors

1. What is the smallest unit of measurement indicated in each of the following measurements? What is the greatest possible error of measurement? Copy the table and fill in the blanks.

Recorded measurement	Smallest unit of measure	Greatest possible error
a. 3 hours 13 minutes		
b. $9\frac{5}{8}$ inches		
c. 12.68 feet		
d. $4\frac{0}{16}$ inches		
e. $156\frac{8}{16}$ cubic feet		

2. What is the tolerance interval for each of the following recorded measurements? Copy the table and fill in the blanks.

Recorded measurement	Upper limit	Lower limit
a. (7.4 ± 0.05) inches		
b. $(16\frac{3}{4} \pm \frac{1}{8})$ inches		
c. (6.93 ± 0.005) centimeters		
d. $(3\frac{0}{8} \pm \frac{1}{16})$ inches		

Accuracy of Measurement

Although two recorded measurements may have the same precision, the size of the absolute error may be more important in one than in the other. An error of $\frac{1}{2}$ inch may not be important in measuring the width of your yard, but it would be serious in building a cupboard in the kitchen. In measuring the distance to the sun, not even an error of 100,000 miles is considered misleading!

Suppose that we make measurements of 8 inches and 80 inches, precise to the nearest inch. The absolute error for both of these measurements is $\frac{1}{2}$ of 1 inch or 0.5 inch. These measurements could be expressed as:

$$(8 \pm 0.5) \text{ inches and}$$
$$(80 \pm 0.5) \text{ inches.}$$

It is clear that the error 0.5 is a larger part of 8 than it is of 80. If we write the ratio (a division comparison) of the absolute error to the recorded measurements, we get:

$$\frac{0.5}{8} = \frac{5}{80} = .0625 = 6.25\%, \text{ and}$$

$$\frac{0.5}{80} = \frac{5}{800} = .00625 = .625\%.$$

A comparison of these per cents shows that the error in the first measurement is more serious than the same error in the second measurement. The above ratio is called the *relative error* of measurement. The relative error is expressed by the following formula:

$$\text{relative error} = \frac{\text{absolute error}}{\text{recorded measurement}}$$

The per cent of error is equal to the relative error multiplied by 100. The accuracy of a measurement is always given in terms of the relative error. The *smaller* the relative error the *greater* the accuracy of the measurement. Do not confuse accuracy with precision. *The precision of a measurement is determined by the absolute error, the accuracy by the relative error.*

EXERCISE SET 12
Precision or Accuracy?

1. Give the absolute error (greatest possible error) for each of these measurements.
    a. 6 feet
    b. 4.8 pounds
    c. $27\frac{1}{2}$ minutes
    d. 0.04 gram
    e. 560 miles
    f. 8.204 meters

2. Find the relative error for each measurement in problem 1.

3. Copy and complete the following table by finding the unit of measure, the absolute error, and the relative error for each of the measurements given.

Measurement	Smallest unit of measure	Absolute error	Relative error
a. 5.6 meters	0.1 meter	0.05 meter	$\dfrac{0.05}{5.6} = 0.0089$
b. 560 meters (to nearest 10 meters)			
c. 0.056 meter			
d. 5,600 meters (to nearest 100 meters)			

4. Which one of the measurements in problem 3 is the most precise? Which is the most accurate?

5. Express each measurement given in problem 3 as the product of the smallest unit of measure and a whole number. For example, in measurement *a* the smallest unit of measure is 0.1 meter. This unit is contained in 5.6 meters 56 times. Thus 5.6 meters equals 56 (a whole number) × 0.1 meter (the smallest unit of measure).
    a. 5.6 meters   = 56 × 0.1 meter
    b. 560 meters   =
    c. 0.056 meter   =
    d. 5,600 meters =

$$3.14 \times 6 = 18.8\cancel{4} \approx 18.8$$
$$3.1416 \times 6 = 18.8\cancel{496} \approx 18.8$$

Significant Digits

In all of the measurements in problem 3 of Exercise Set 12, the relative error, or accuracy, is the same. Problem 5 shows that, although the basic unit of measure is different for each measurement, the number of units in each measurement is always the same, 56. For each measurement, then, there are two digits of significance, the 5 and the 6. These digits are considered to be significant because they can be used to show the accuracy of a measurement. The more significant digits in a measurement, the more accurate the measurement. The accuracy of a measurement is independent of the unit of measure used to make the measurement.

Let's see how we find the significant digits in a numeral so we can quickly determine the accuracy of a measurement.

Significant digits can be defined as digits which affect the accuracy of a measurement. The way to determine whether a digit is significant is to see if it is used to tell us how many measured units are involved. This is illustrated in Table 5.

Table 5

Recorded measurement	Smallest unit of measure	Number of measured units	Number of significant digits
475 in.	1 in.	475	3
5,600 ft. (to nearest 100 ft.)	100 ft.	56	2
2.86 cm.	0.01 cm.	286	3
8.056 sec.	0.001 sec.	8,056	4
0.0005 gm.	0.0001 gm.	5	1
93.0 in.	0.1 in.	930	3
0.0670 kg.	0.0001 kg.	670	3

By finding the basic unit of measure and the number of times that unit is used, we can determine the significant digits of the number expressing the measurement. From Table 5, the following rules for significant digits become evident.

All non-zero digits are significant. Each of these numerals has three significant figures: 578, 23.2, .754, 8.26. Each digit is significant in telling us the units of measure involved.

Any zero that is between non-zero digits is significant. The zeros are significant in each of these four-significant-digit numerals: 3.075, 560.8, 9.043, 67.05. Here the zeros count groups that are necessary in giving us the total number of measurement units.

Final zeros of a number expressed as a decimal fraction are always significant. These numbers have three significant figures: 0.540, 32.0, 80.0, 0.0600. Check by finding the number of units of measure indicated by each number.

Zeros standing alone at the left of a number expressed as a decimal fraction are never significant. The zeros in these numbers are not significant: 0.07, 0.005, 0.468, 0.0001.

Final zeros of a whole number may or may not be significant. In a number such as 5,600 it is difficult to know what the unit of measurement is. If the number is 5,600 to the nearest hundred, then the zeros are not significant. If the measurement is correct to the nearest unit, then both zeros are significant. To avoid misunderstanding in a situation like this, a dot can be placed over each zero which is significant, like this — 5,6̈0̈0. Another way to show the significant digits is to write the number in scientific notation by writing it as the product of a number between 1 and 10 and a power of 10. Then, $5,6\ddot{0}0 = 5.600 \times 10^3$ and $5,600 = 5.6 \times 10^3$. 5.600 indicates four significant digits, while 5.6 indicates two.

The following examples illustrate the definition of significant figures. The number of significant digits in each case is shown in parentheses. Note carefully the zeros which are significant.

374	(3)	0.0374	(3)
0.00374	(3)	374.0	(4)
37,400	(3)	372.040	(6)
3.7400	(5)	374,000	(3)
0.003740	(4)	3,740.0	(5)
3,074	(4)	37,4̈00	(4)

The number of significant figures in fractional expressions is determined by counting the digits in the numerator. A mixed number must first be changed to an improper fraction. For example, $9\frac{5}{12}$ has three significant digits, since $9\frac{5}{12} = \frac{113}{12}$. When more than one unit of measurement is involved, the measurement must be expressed in terms of the smallest unit before counting the significant figures. For example, 9 pounds 3 ounces equals 147 ounces and has three significant figures. These rules are consistent in terms of the unit of measure involved.

EXERCISE SET 13
Determining Significant Figures

1. How many significant figures does each of the following measured quantities have? Copy and complete the table.

Recorded measurement	Unit of measure	Number of units involved	Number of significant figures
a. 5.72 ft.	0.01 ft.		
b. 0.068 in.	0.001 in.		
c. 137.0 sec.	0.1 sec.		
d. 0.009 cm.	0.001 cm.		

2. How many significant figures does each of these numerals have? Watch out for significant zeros.

a. 507

b. 920

c. 80.6

d. 0.0580

e. 7,300

f. 104.0

g. 62.005

5,849,713,600,000   5.8 Trillion

Rounding Off Numbers

When we compute with measured numbers, we often round off the numbers so that they will show the precision or accuracy that is appropriate. In rounding off numbers, we drop digits or replace digits with zeros to make a numeral easier to use and interpret. Instead of saying 65,128 people attended the game last Saturday, we would probably round the value off to 65,000 people. When we replace digits with zeros by rounding off, the zeros are not significant. Make the numbers in the following statements simpler by rounding them off to two significant figures.

The satellite traveled 75,281 miles away from the earth.

The game commission placed 4,673 fingerling bass in Long Lake.

The length of our playground is 324.3 feet.

The time needed to travel home from downtown was 28.5 minutes.

Since many "authorities" differ on how to round off numbers, we will use the following rules for rounding numbers in this booklet.

In rounding off a number, the digits dropped must be replaced by "place-holding" zeros. If the first of the digits to be dropped (reading from left to right) is 1, 2, 3, or 4, simply replace all dropped digits with the appropriate numbers of zeros. Thus, 57,384 rounded off to the nearest thousand becomes 57,000.

If the first of the digits to be dropped (reading from left to right) is 6, 7, 8, or 9, increase the preceding digit by 1. For instance, 5,383 rounded off to the nearest hundred becomes 5,400.

If only one digit is to be dropped and this digit is 5, increase the preceding digit by 1 if it is odd, and leave it unchanged if it is even. If more than one digit is to be dropped and the first digit is 5, increase the preceding digit by 1. Thus, if 785 is to be rounded off to the nearest ten it becomes 780, while 635 rounded off to the nearest ten becomes 640.

If a decimal fraction is rounded, the digits which are to the right of the decimal point should not be replaced by zeros, because zeros to the right of a decimal are significant. For example, 73.2 rounded off to one significant figure becomes 70., not 70.0.

EXERCISE SET 14
Rounding Numbers

Round off these numbers to the stated number of significant figures.

	Numeral	Round off to this number of significant figures
1.	8,465	3
2.	0.7394	2
3.	473,596	3
4.	84.23	2
5.	245,086	4
6.	406.50	3

Addition and Subtraction with Measurements

What is the sum of the measurements 5.2 inches and 2 inches? 7.2 inches would not be the proper answer. When we compute with most numbers arising from measurements, we need new rules. This is necessary because measurements obtained with measuring devices are never exact. If our measurements are approximations, then our computations with these measurements are also approximate. It is impossible to increase the precision or accuracy of a measurement by computation.

Let's look at the measurements we mentioned earlier. The measurement 5.2 inches indicates that it is made to the nearest tenth of an inch. Thus, the actual length may be as small as 5.15 inches or as large as 5.25. Similarly, the measurement 2 inches may be as small as 1.5 inches or as large as 2.5 inches. If we compute the sum of the largest possible values and of the smallest possible values, we will find a considerable difference in the result.

Smallest possible measures	Largest possible measures
5.15 inches	5.25 inches
+ 1.5 inches	+ 2.5 inches
6.65 inches	7.75 inches

Thus, the sum of the two measurements *must* be some value in the interval 6.65 inches to 7.75 inches. The chances are very small that the sum of 5.2 and 2 inches is exactly 7.2 inches. If we wrote the sum as 7.2 inches, we would be indicating a possible error of only 0.05 inch, or, in other words, we would be indicating that the sum might be between 7.15 and 7.25 inches. However, we have seen that the result actually might be any value from 6.65 to 7.75 inches. Since we can't increase the precision of a measurement by computation, it seems reasonable to give the sum the same precision as the least precise of the original measurements, 2 inches. Therefore, we round off the sum of 5.2 inches and 2 inches to 7 inches. The precision of this answer is expressed by a possible error of 0.5 inches, which suggests an interval of 6.5 inches to 7.5 inches for the exact value of the sum. This is much closer to the 6.65- to 7.75-inch possible range we discovered above. In many cases, the errors in the original measurements would cancel each other to some extent. Then the possible error in the sum of the measurements would probably be no greater than the possible error of the least precise measurement.

This discussion brings out the following general principle for adding (or subtracting) measurements. *The sum (or difference) of measures should be rounded off to the unit of the least precise of the measurements involved.* For example, the sum of $3\frac{5}{8}$ inches and $2\frac{1}{2}$ inches is $6\frac{1}{8}$ inches, which should be rounded to the nearest $\frac{1}{2}$ inch, or $6\frac{0}{2}$ inches. We keep the fraction $\frac{0}{2}$ to show the precision of the result.

Sometimes sums of measurements expressed in the United States system can be simplified by changing to larger basic units.

In some subtraction problems, such as example **C** below, it is necessary to change to smaller units to make subtraction possible.

Examples:

A. Add:

5 ft. 5 in.

+ 4 ft. 7 in.

9 ft. 12 in. or 10 ft. 0 in.

(Why do we keep the zero?)

B. Add:

5.48 meters

+ 7.2 meters

12.68 meters or

12.7 meters

C. Subtract:

7 lb. 2 oz. = 6 lb. 18 oz.

− 3 lb. 9 oz. = 3 lb. 9 oz.

3 lb. 9 oz.

D. Subtract:

$5\frac{7}{8}$ in. = $5\frac{7}{8}$ in.

$-2\frac{1}{2}$ in. = $2\frac{4}{8}$ in.

$3\frac{3}{8}$ in. or $3\frac{1}{2}$ in.

EXERCISE SET 15
Adding and Subtracting Measurements

1. Add the following measures and round each sum to the proper precision.

a.   3.568 sec.

+ 5.2   sec.

b. $2\frac{3}{8}$ in.

+ $5\frac{1}{2}$ in.

c.   4 lb. 9 oz.

+ 3 lb. 7 oz.

d. $3\frac{5}{8}$ in.

$4\frac{1}{4}$ in.

$2\frac{0}{2}$ in.

e. 6   grams

3.4 grams

5.4 grams

f.   8 cm. 6 mm.

13 cm. 9 mm.

4 cm.

2. Perform the subtractions involving the following measures and round each difference to the proper precision.

a.   4.57 liters

− 1.6   liters

b. $17\frac{1}{2}$ in.

− $9\frac{5}{8}$ in.

c.   3 gal. 1 qt.

− 1 gal. 2 qt.

d. $7\frac{2}{3}$ ft.   e.   4   grams   f.   6 meters 24 cm.

    $- 2\frac{0}{2}$ ft. $- 2.8$ grams $- 2$ meters 18 mm.

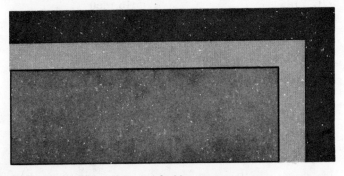

Multiplication and Division with Measurements

The senior boys plan to set up a front for a booth at the school carnival by putting three tables end to end. If each table is 4.3 feet long, the total length of the front of the booth can be computed by multiplying 4.3 feet by 3. The 3 is an exact number, for it resulted from a count. This problem could be written as

$$3 \times 4.3 \text{ feet} = 12.9 \text{ feet.}$$

This multiplication means a repeated addition of 4.3 feet 3 times. Therefore, we can use the law for addition of measurements and give the product the same precision as the original measurement.

The multiplication of one measurement by another is a different story. In multiplication, recall that we multiply one number by another number. In measurement, then, we multiply the numbers of the measurements together and associate this product with the correct unit. For example, if a rectangle has the dimensions 4 inches and 3 inches, the area is (4×3) square inches. We can illustrate some of the difficulties that come up in such a multiplication by working with a rectangle. Suppose that the dimensions of a rectangle are measured as 4.2 inches and 2.6 inches. These dimensions indicate that the smallest unit of measure is 0.1 inch, with an absolute error of 0.05 inch. The tolerance interval for the dimensions can be given as (4.2 ± 0.05) inches and (2.6 ± 0.05) inches. The pair of longest possible dimensions would be 4.25 and 2.65 inches and the pair of the smallest possible dimensions would be 4.15 and 2.55 inches. Figure 9 illustrates how the rectangle might vary in size according to these dimensions.

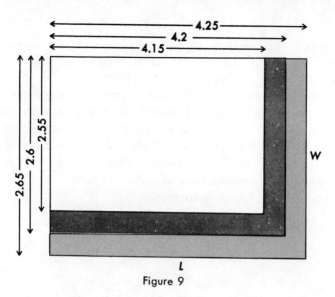

Figure 9

An indirect measurement of the area of the rectangle can be made by computing the product of the measures of the two dimensions ($A = lw$). The measure of the area of our rectangle might be represented by any one of the following numbers or any number between them:

$A_1 = (4.15 \times 2.55)$ sq. in. $= 10.5825$ sq. in. (smallest possible area)

$A_2 = (4.2 \times 2.6)$ sq. in. $= 10.92$ sq. in.

$A_3 = (4.25 \times 2.65)$ sq. in. $= 11.2625$ sq. in. (largest possible area)

If we had to pick one value of the three just listed for the area, we would probably pick the middle result, 10.92 square inches. But if we try to compare the *precision* of this result with the *precision* of the dimensions, we run into difficulty. We cannot compare the precision of a linear measure with the precision of an area measure. Only *one* measurement is made for a length, but *two* measurements are made for an area. Since we multiply the two measurements to find the area, there is a chance that our error of measurement will be multiplied, too. This is illustrated in the above example.

Since it is impossible to make a *precision* comparison between the product and the factors of the product, let us compare the accuracies of the measures. If we round the numbers 10.5825, 10.92, and 11.2625 to two *significant figures*, we obtain 11 as a result in each case. Rounding to two significant figures gives a common answer. We should also note that the original dimensions contained two significant digits. Hence, we get an acceptable answer for our product of two measurements by making the result agree approximately with the starting values in *accuracy* (by having the same number of significant digits).

Another way to illustrate multiplication with measured numbers is to keep track of the numerals which repesent approximations. Suppose we wish to find the product of 4.3 inches and 1.52 inches. In the computation, the 3 and 2 are shaded to show that these digits represent approximations. The products obtained from the 3 and the 2 and the sum of these products are also shaded, for they will be approximate, too.

$$
\begin{array}{r}
1.5\,2 \\
\times\ 4.3 \\
\hline
4\ 5\ 6 \\
6\ 0\ 8 \\
\hline
6.5\ 3\ 6
\end{array}
$$

In the product the 5, 3, and last 6 are all approximations. This product should be rounded off to 6.5 square inches, since no more than one digit representing approximations is usually retained in a recorded measure. This result, then, has two significant figures, which is the same number as in the least accurate factor of the product. These examples illustrate the following rule for multiplication of inexact quantities: *when multiplying two inexact quantities, the product should contain the same number of significant digits as the least accurate factor of the product.*

Let's consider a different situation in which it will be necessary to find the product of a measurement and another type of approximation. Suppose Mark wants to find out how far his bicycle wheel travels in one revolution. To find this, he must compute the circumference of his 28-inch-diameter bicycle wheel. He knows the formula for the circumference is $C = \pi d$ and that π has an approximate value of 3.1416. He finds the circumference of the wheel by doing the following work.

$$
\begin{array}{r}
3.1416 \\
\times\ 28 \\
\hline
251328 \\
62832 \\
\hline
87.9648
\end{array}
$$

If Mark applies our rule for multiplication of numbers representing inexact quantities, he will round off the results to two significant digits and accept 88 as an answer. By examining the following multiplications, let's see how Mark could have saved some work in obtaining this answer.

$$
\begin{array}{cc}
3.14\text{16} & 3.14 \\
\underline{28} & \underline{28} \\
2513\text{28} & 2512 \\
\underline{628\text{32}} & \underline{628} \\
87.9\text{648 or 88} & 87.92 \text{ or } 88
\end{array}
$$

We can see that the "16" in the first multiplication has no effect on the two significant digits retained in the answer, 88. Therefore, the same result is obtained by doing the easier multiplication of 3.14 and 28. *When multiplying two numbers representing approximations, one factor should have, at most, one more significant digit than the other.*

Perhaps we should also note in the multiplication of 3.14 and 28 that, since the 8 represents an inexact value, all digits in the final result represent approximations. Hence, in the answer, 88, both digits represent approximations. Nevertheless, it is still probably better to make the product agree with the accuracy (as indicated by significant digits) of the least accurate factor instead of rounding to one significant digit.

The rules for division follow from multiplication, since division is the inverse of multiplication. If 3 feet 5 inches times 4 (where 4 is the result of a count) equals 12 feet 20 inches or 13 feet 8 inches, then 13 feet 8 inches divided by 4 must equal 3 feet 5 inches.

$$
\begin{array}{r}
3 \text{ ft.} \qquad 5 \text{ in.} \\
\hline
4\overline{)13 \text{ ft. } 8 \text{ in.}} \\
12 \text{ ft.} \\
\hline
1 \text{ ft. } 8 \text{ in.} = 20 \text{ in.} \\
\underline{20 \text{ in.}}
\end{array}
$$

In this division 4 is considered an exact number.

However, 3.15 feet divided by .5 feet becomes 6.3, which should be rounded off to 6, one significant figure. *When dividing numbers obtained by measurements, keep as many significant digits in the quotient as there are in the measurement with the fewest significant digits.* Note that, when one measurement is divided by another measurement with the same units, the quotient is not expressed as a measurement.

A similar rounding rule applies to square root. For example, if a square contains 75 square inches measured to the nearest square inch, the accuracy is indicated by two significant digits. The side of this square is $\sqrt{75}$ inches or 8.5 inches, a length correct to two significant digits.

EXERCISE SET 16
Multiplication and Division Computations

Perform the indicated computations and round to the proper accuracy.

1. Multiply.

a.   2.3 cm.   b. $48\frac{1}{2}$ in.   c.   15 ft.   d. $(0.4 \text{ in.})^3$

       $\times 4$ $\times 4\frac{1}{8}$ in. $\times 0.03$ ft.

2. Divide.

a. $\dfrac{9 \text{ ft. } 3 \text{ in.}}{4}$ b. $\dfrac{6.4 \text{ cm.}}{0.2 \text{ cm.}}$ c. $\dfrac{7\frac{3}{8} \text{ in.}}{5 \text{ in.}}$ d. $\dfrac{186.4 \text{ yd.}}{14 \text{ yd.}}$

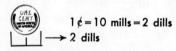

1 ¢ = 10 mills = 2 dills

⟶ 2 dills

Converting from One Unit to Another

Some of the most troublesome problems in measurement work arise when it is necessary to change from one unit of measure to another. Let's examine some problems of this nature by setting up a new system of measurement.

Any size units can be invented for measurements of quantities such as length or weight. Let's let the following statements describe the relationship between the units of a new system for length measurement.

> one pill = 5 mills
> one dill = 5 pills
> one will = 5 dills

What length is represented by a mill? Naturally, we don't know until it is described. Let's say that 10 mills have a length equal to the diameter of a penny.

Can you use this table to find how many "pills" are in one "will"? As soon as you know how two measures compare in size, you can change from one unit to another. If you are changing "pills" to "mills", do you multiply by 5 or divide by 5? If you are changing "dills" to "wills", do you multiply by 5 or divide by 5? If we are changing from a large unit to a smaller unit, we multiply by the number that tells us the number of smaller units equivalent to one large unit. If we are changing from a small unit to a larger unit, we have to divide by the number that tells us the number of small units in each larger unit.

Another way to remember how to change from one unit of measure to another is by working with rates (a comparison by division of two measurements expressed in different units). For example, the rate of "pills" to "mills" can be expressed as $\dfrac{1 \text{ pill}}{5 \text{ mills}}$. The rate of "pills" to "mills" will be constant, always equal to $\dfrac{1 \text{ pill}}{5 \text{ mills}}$. Hence, if we wish to find the number of "pills" equivalent to 145 "mills", we can write the following equality statement:

$$\frac{1 \text{ pill}}{5 \text{ mills}} = \frac{x \text{ pills}}{145 \text{ mills}}$$

$$\text{or } \frac{1}{5} = \frac{x}{145},$$

where x holds a place for the number of "pills."

Solving this equality statement we have

$$x = \frac{145}{5} \text{ or } x = 29 \text{ pills.}$$

EXERCISE SET 17
Working with New Units of Measure

Here is a new system of units for time.

one day = 20 rays
one ray = 20 says
one say = 20 ways

Answer the following questions involving these units and the units of length presented in the preceding section.

1. Do the following conversions using the new units described above.
  a. 1,000 ways = ? rays   b. 35 pills = ? mills
  c. 95 dills = ? wills   d. 90 rays = ? says

2. Compute as indicated with the above time units.

Add	*Subtract*
a. 7 rays 14 says	b. 13 rays 12 says
+ 5 rays 7 says	− 4 rays 17 says

Multiply	*Divide*
c. 6 says 9 rays	d. 13 says 11 ways
× 5	2

3. Compute as indicated with the above length units.

	Add		*Subtract*
a.	3 dills 4 pills	b.	2 wills 3 dills
	+ 4 dills 2 pills		− 1 will 4 dills

	Multiply		*Divide*
c.	4 pills 3 mills	d.	4 dills 1 pill
	× 2 mills		3

$$3.14159+$$

$$\sqrt{3} = 1.732$$

$$\left(1 + \frac{1}{N} \right)^{N} = e = 2.7182818284+$$

$$\pi = \frac{c}{d}$$

$$\sqrt{2} = 1.414+$$

The Mathematics of Approximation

The principles for computing with measurements apply to all approximations. And approximations occur from many sources other than measurement. For example, approximations arise when we attempt to write a decimal numeral for a number like π, $\frac{1}{7}$, or $\sqrt{3}$. Numbers such as 93,000,000, obtained by rounding to the nearest million, also represent approximations. Remember that the number itself is not approximate. However, when we use the numeral 93,000,000 to represent the approximate distance to the sun in miles, we understand that the 93,000,000 is an approximation of the true number of miles to the sun.

The representation of $\frac{1}{3}$ in a decimal form provides a simple example of approximation that is not related to measurement. By carrying out the indicated division, we know that $\frac{1}{3}$ can be written as 0.33, but we also know that $\frac{1}{3}$ is not exactly equal to 0.33. However, we know that $\frac{1}{3}$ is greater than 0.33 but less than 0.34 and, hence, is between 0.33 and 0.34. Since we know that the difference between $\frac{1}{3}$ and 0.33 is less than the difference between 0.34 and $\frac{1}{3}$, we say that $\frac{1}{3}$ is approximately equal to 0.33.

Some of these ideas can be symbolized in the following way.

$\frac{1}{3}$ is greater than 0.33   or $\frac{1}{3} > 0.33.$

$\frac{1}{3}$ is less than 0.34   or $\frac{1}{3} < 0.34.$

$\frac{1}{3}$ is between 0.33 and 0.34   or $0.33 < \frac{1}{3} < 0.34.$

$\frac{1}{3}$ is approximately 0.33   or $\frac{1}{3} \approx 0.33.$

Of course, we can get a better approximation for $\frac{1}{3}$ by writing 0.333, or 0.3333, or 0.33333, and so on, each of which is a closer approximation to $\frac{1}{3}$ but none of which exactly equals $\frac{1}{3}$.

We can use the above symbolism when working with measurements. For example, the statement 1 inch = 2.54 centimeters could be written as

$$1 \text{ in.} \approx 2.54 \text{ cm.,}$$

indicating

$$2.535 \text{ cm.} < 1 \text{ in.} < 2.545 \text{ cm.}$$

Notice that we use the symbol $\approx$ to represent "is approximately equal to." We can also use graphs to picture approximations. Suppose we know that some quantity, represented by x, is approximately 3. If

$$x \approx 3$$

is interpreted as meaning

$$2.5 < x < 3.5,$$

then this approximation could be pictured on a number line as shown in Figure 10.

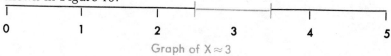

Graph of X ≈ 3

Figure 10

All numbers between 2.5 and 3.5 are represented by points on the graph (in gray) in Figure 10. The graph indicates that any number between 2.5 and 3.5 would be considered as approximately equal to 3.

We can also picture approximation relationships between *two* variable quantities represented as x and y. If we interpret $x \approx y$ as meaning $x - y < \frac{1}{2}$ or $y - x < \frac{1}{2}$ (that is, two quantities are approximately equal if the difference between the two quantities is less than $\frac{1}{2}$), then the graph of $x \approx y$ is like the one shown in Figure 11.

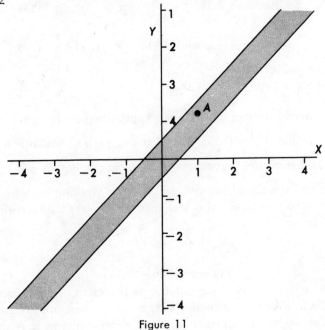

Figure 11

Every point in the shaded area of Figure 11 represents a *pair* of numbers that satisfy the relationship $x \approx y$. For example, point A represents the pair of numbers $x = 1$, $y = 1\frac{1}{4}$. In this situation, we would interpret these numbers as approximately equal since they differ by less than $\frac{1}{2}$.

EXERCISE SET 18
Approximations

1. Use symbols to express the following statements.
   a. An approximation for π is 3.1.
   b. A decimal approximation for $\frac{2}{3}$ is between 0.66 and 0.67.
   c. The sum of two quantities (x and y) is approximately equal to 2.

2. Use a number line to graph the following approximations.

a. $x \approx 5 \ (4.5 < x < 5.5)$

b. $x \approx \dfrac{1}{2} \ (\dfrac{1}{4} < x < \dfrac{3}{4})$

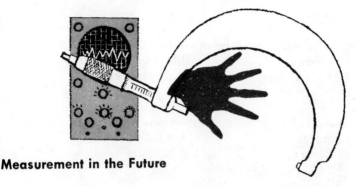

Measurement in the Future

There have been three big steps in improving measurement during the last 500 years. The first occurred when mankind realized that measures using parts of the human body were inaccurate. To correct this, standards independent of a man's size were established. The second step occurred when machines such as the micrometer, which measured more precisely, were made. The third step is now underway. Scientists and mathematicians are inventing machines that use electrons, light waves, sound waves, and radio waves as the yardsticks of tomorrow. These machines will make measurements that we cannot perceive with our limited senses of sight, hearing, or feeling.

Of course, man's ability to perceive is limited and, in many ways, is not as great as that of animals. For example, the reindeer scents the wolf miles distant, the eagle sees a salmon leap a mile away, the spider works with webs too light to feel, a bird navigates a 6,000-mile trip without any sort of device for measuring distance or direction. Man has had to use his intelligence to invent instruments to improve his perception. We have sound instruments that catch and measure the noise of molecules, spectroscopes that measure the make-up of distant stars, scales that weigh a speck of dust, and radar devices that navigate a satellite in space.

As the Greeks learned to measure the earth's circumference long before Columbus sailed westward, so we must make measurements of new quantities on our earth and in space before new and exciting knowledge can be found. A ladder is needed to climb to new heights. In the fields of science and mathematics, many of the rungs of this ladder are measuring instruments that can give necessary information for a new advance into the unknown.

EXERCISE SET 19
Measurement Review

1. What is the unit of measurement in each of the following?

    a. 18.6 inches

    b. 4.03 feet

    c. $5\frac{5}{8}$ inches

    d. 6 pounds 9 ounces

    e. 0.008 second

2. What is the absolute error of measurement in each of the following?

    a. 15 minutes

    b. 9.0 ounces

    c. $6\frac{3}{4}$ inches

    d. 6,000 miles

    e. 9.42 feet

3. Round off the following measurements so that all will have the same degree of precision.

    a. 6.94 inches

    b. 10.605 inches

    c. 0.0057 inch

    d. 56.00 inches

    e. 3.995 inches

4. Round off the following measurements so that all will have the same degree of accuracy as the least accurate measurement.

    a. 468.5 meters

    b. 0.00708 meter

    c. 3.467 meters

    d. 5693 meters

    e. 3.004 meters

5. Arrange the following measurements in order of precision, beginning with the most precise.

    a. 17.04 inches

    b. $5\frac{3}{4}$ inches

    c. 843 inches

    d. 0.006 inch

    e. 3420 inches

6. Arrange the following measurements in order of accuracy, beginning with the most accurate.

    a. 0.0056 meter

    b. 24.3 meters

    c. $\frac{7}{8}$ meter

    d. 14,296 meters

    e. 7.002 meters

7. Indicate the number of significant figures in each of the following numbers.

    a. 25,070

    b. 0.006804

    c. 12.50

    d. 6.03×10^5

    e. 3 feet $6\frac{1}{2}$ inches

    f. $2\frac{1}{3}$ feet

8. Compute the following approximations. Round off each answer to the proper degree of accuracy.

a. $4.08 + 16.7$

b. $3\frac{1}{2} + 4\frac{3}{8}$

c. $8.72 - 2.8$

d. $14\frac{5}{16} - 7\frac{5}{16}$

9. Multiply these approximations. Round off each answer to the proper degree of accuracy.

a. $\begin{array}{r} 4.63 \\ \times\ 5.8 \\ \hline \end{array}$

b. $\begin{array}{r} 74 \\ \times\ 0.064 \\ \hline \end{array}$

c. $\begin{array}{r} 8\frac{3}{4} \\ \times\ 4 \\ \hline \end{array}$

d. $(0.8)^2$

10. Perform the indicated operations with these approximations. Round off each answer to the proper degree of accuracy.

a. $\dfrac{6.852}{12}$

b. $\dfrac{10.75}{5.0}$

c. $3\frac{1}{4} \div \frac{1}{8}$

d. $\sqrt{81.00}$

11. Which of the following statements are always true?

a. Zeros are always significant.

b. Zeros only indicate place value and are, therefore, never significant.

c. Zeros to the left of the decimal point are never significant.

d. Zeros between significant digits are always significant.

e. Zeros to the right of the decimal point are always significant.

12. Copy and complete the following table.

Recorded measurement	Implied unit of measure	Absolute error	Relative error	Significant figures
a. 31.4 inches	0.1 inch	0.05 inch	$\dfrac{0.05}{31.4} = 0.0016$	3
b. 8.32 seconds				
c. $9\frac{3}{8}$ miles				
d. 80 pounds				
e. 18.0 quarts				
f. 5000 cubic feet				
g. 0.0050 millimeter				
h. 1.008 cubic centimeters				

13. The earth is about 93,000,000 miles from the sun. If the path around the sun is assumed to be circular, about how far do you travel around the sun in one year? In one day? In one hour?

Measurement Projects

Here are some measurement activities that would be good projects for a report to your mathematics class.

1. Make a gauge for measuring thickness to the nearest hundredth of an inch. On a piece of cardboard or plywood, lay off line m and p perpendicular to an edge at points A and B, with m and p exactly one inch apart. Take a strip of graph paper with 100 equally spaced divisions and place one end at point A. Mark point C where the strip intersects line p. Cut out the section ABC. Glue the strip of graph paper along edge AC. When a piece of material is placed in the slot, if it touches AC at the 48th mark from C, it is .48 inch thick.

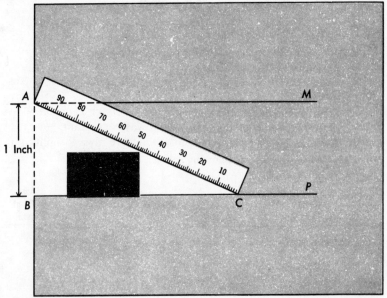

Figure 12

2. Make an exhibit on measurement. Collect and label with sizes such things as nails, screws, bolts, pins, needles, tin cans, gloves, thread, wire, tape, pipe, hats, shoes, paper, and so on.

3. Make a simple micrometer with an ordinary bolt. Measure the distance between threads by counting the threads along one inch of length and dividing that number into 1. Screw a nut tight to the bolt. When you unscrew the nut one complete turn, the space between the nut and the top of the bolt will be the distance between two consecutive threads. Counting the number of turns

needed before a board can be fitted between the top of the bolt and the nut will enable you to determine the thickness of the board.

Attach a pointer to the nut. This pointer will go around as the nut unscrews. Attach a dial behind this pointer to tell you what part of a turn has been completed. If this dial is marked into ten equal arcs, each unit on the dial will show $\frac{1}{10}$ of the width between threads. Attaching the bolt and dial to a board makes the device more convenient to use. An engineer's micrometer uses a screw with 40 turns to the inch and a dial divided into 25 parts.

4. Create a model to measure the tensile strength of thread. How could you measure the weight a string will support before it breaks?

5. Find the weight a floating object will support by experimenting with empty cans of different sizes. Float these cans in water in a laundry tub. Add stones until the cans barely float. Weigh the stones in each can. Find the weight of each can full of water. How do these weights compare?

6. Visit engineers, carpenters, bricklayers, science laboratories, airports, weather bureaus, government agencies, shops, or factories to find out the different kinds of measurements used.

7. Make a bulletin-board display of different measuring instruments and units of measure, new and old.

8. Learn to use a transit and measure the height of your school or map the playground.

9. Collect measurement data about sports records such as the highest jump, the fastest race, the longest home run, or the biggest fish ever caught.

10. Find the rate of travel of insects, birds, animals, and fishes. Make a graph or chart of these rates to show how they compare.

11. Make a progress chart of the speed at which man has been able to travel in different ages by walking, swimming, riding on different animals, boating, automobiles, trains, airplanes, and rocket flights.

12. Lay off a distance of 52.8 feet or 528 feet. Measure the time needed to walk the distance. Since the first distance is $\frac{1}{100}$ of a mile and the second $\frac{1}{10}$, it will be easy to compute the time needed to walk a mile. With this information you can get a measure of walking speed in miles per hour.

13. Find the meanings of the sizes of these articles.

    a. 20-penny nail

    b. square of roofing

    c. $\frac{3}{8} \times 1$ screw

    d. $2\frac{1}{2}$ pencil

    e. 16-pound paper

    f. 75-watt light

    g. number two can

    h. $\frac{3}{4}$-ton air conditioner

    i. building brick unit

    j. $\frac{3}{4}$-inch pipe

    k. 50,000-B.T.U. (British thermal unit) furnace

    l. cord of wood

14. Find recorded measurements of common objects or distances and determine the absolute error and relative error — for example, distance to sun, diameter of a hair, weight of an atom, and so on.

Extending Your Knowledge

If you have enjoyed exploring the measurement ideas presented in this part of the book, you may want to consult some of the following publications for further information.

Selected Pamphlets

The Amazing Story of Measurement. Lufkin Rule Co., Saginaw, Michigan

How Long Is a Rod? Ford Motor Co., Dearborn, Michigan

Precision, a Measure of Progress. General Motors, Detroit 2, Michigan

PALMER, E. LAWRENCE, *Let's Measure Things.* Cornell University, Ithaca, New York

SHUSTER, CARL, *Computation with Approximate Data.* Yoder Instrument Co., East Palestine, Ohio

SWEINHART, JAMES, *He Measured in Millionths.* Ford Motor Co., Dearborn, Michigan

Books

BENDICK, JEANNE, *How Much and How Many.* McGraw-Hill Book Co., 1947

RASSWEILER, M. and HARVIS, J. M., *Mathematics and Measurement.* Row, Peterson and Co., 1955

NEUMAN, JAMES, *The World of Mathematics.* pp. 1797-1813, Simon and Schuster, 1956

SAWYER, W. W. and STRAWLEY, L. G., *Designing and Making.* Basil Blackwell, Oxford, 1952

Charts

Dimensions of Natural Objects. Central Scientific Co.

International Metric System. Welch Scientific Co.

PART III

ADVENTURES
in Graphing

RED FEATHER TIME

Graphing: A Story
of Mathematics in Pictures
LET'S GO OVER THE TOP

Numbers and Lines

Suppose that your class is having a campaign to collect money for charities in your town and that you are in charge of recording progress made in the campaign. You might want to make a visual aid to show the progress made toward reaching the goal set by the class. This could be done by drawing a line segment on a piece of paper, drawing evenly spaced marks on the line segment, and then labeling the marks from 0 up to the number of dollars set as a goal. A numbered line like this is shown in Figure 1.

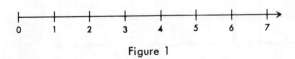

Figure 1

The progress made in the drive could be shown each day by marking, in some way, the number representing the total amount of money collected up to that time.

In the above example, certain numbers were represented by marking points on a line. *It would theoretically be possible to represent any number with a point on a line.*

The number-line idea is also illustrated by a common thermometer like the one shown in Figure 2.

Figure 2

The thermometer is merely a sealed glass tube which contains a liquid and on which a number line has been drawn. As the liquid in the thermometer is warmed or cooled, it expands or contracts. Thus, the top edge of the liquid might be at various positions with respect to the number line on the thermometer. Nevertheless, *any point on the thermometer scale can be represented by a number*. Of course, thermometer scales are usually so small that the readings are rounded off to the nearest one-half unit.

The above examples illustrate an important principle. We have seen that any number can be compared to a point on a line and that any point on a line can be compared to a number. We say that there is a *one-to-one correspondence* between the set of all numbers and the points on a line. These ideas are quite simple, but very important, for they enable us to use drawings to picture numbers and number relationships. You perhaps recall that a number line can be used to show a simple sum such as $3 + 2 = 5$. Figure 3 illustrates this sum and also illustrates the fact that $3 + 2 = 2 + 3$ (the commutative principle of addition).

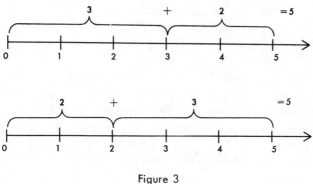

Figure 3

A Number-Pair Game

Let's examine some mathematical ideas that can be illustrated by a pair of dice. Suppose you have one white die and one red die. If you throw the dice, you will obtain a pair of numbers — one from the white die and one from the red die. Table 1 lists all the number combinations that could be obtained by throwing the dice. The number from the white die is listed first, and that from the red die second.

Table 1

1,6	2,6	3,6	4,6	5,6	6,6
1,5	2,5	3,5	4,5	5,5	6,5
1,4	2,4	3,4	4,4	5,4	6,4
1,3	2,3	3,3	4,3	5,3	6,3
1,2	2,2	3,2	4,2	5,2	6,2
1,1	2,1	3,1	4,1	5,1	6,1

In Figure 4, each pair of numbers has been represented by a dot, with the dots being equally spaced.

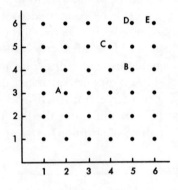

Figure 4

The horizontal rows have been labeled from 1 to 6 to represent the white die numbers. The vertical columns have also been labeled from 1 to 6 to represent the red die numbers. *Each dot in Figure 4 represents a pair of numbers.* For example, dot A represents the number pair (2,3) — 2 on the white die, 3 on the red die. Dot B represents the number pair (5,4) — 5 on the white die, 4 on the red die.

The arrangement of points in Figure 4 is called a *rectangular array*. This arrangement can be used to play a simple number-pair game. Each player is given three markers and the following rules are used:

1. The dice are thrown, and the player who throws the highest total plays first; the one with the second highest plays second, and so on.

2. The first player rolls the dice and places a marker on the dot representing the number combination shown by the dice. Then the second player follows the same procedure, and so on.

3. The object of the game is to move each counter to the (6,6) position and then off the array.

4. When a player rolls the dice after his first turn, the number combination obtained can be applied to a marker on the board by moving it the indicated number of white and red spaces, or it can be applied to a new marker.

5. A marker can be moved off the array only from the (6,6) position.

6. A number combination cannot be applied if it moves a marker off the array from a point other than the (6,6) position.

7. Once a marker is in the (6,6) position, any throw of the dice can be used to move it off the array.

8. Only one marker may be placed on a given position on the array.

9. If a player cannot move any of his markers, he loses his turn.

Let's consider an example. If the first player throws the combination white 4, red 5, he would place a marker on point C shown in Figure 4. On his second play, a throw of $(1,1)$ or $(2,1)$ would permit him to move the first marker to point D or E. If he has any other combination, he will have to use a second marker. If the position he rolls is occupied by another player's marker, he loses his turn.

The concept of an array of points representing a set of number pairs is an important one in mathematics. Let's consider another rectangular array situation by working with all the possible number pairs that can be formed from the numbers -3, -2, -1, 0, 1, 2, 3. A list of these number pairs is given in Table 2.

Table 2

$-3,3$	$-2,3$	$-1,3$	$0,3$	$1,3$	$2,3$	$3,3$
$-3,2$	$-2,2$	$-1,2$	$0,2$	$1,2$	$2,2$	$3,2$
$-3,1$	$-2,1$	$-1,1$	$0,1$	$1,1$	$2,1$	$3,1$
$-3,0$	$-2,0$	$-1,0$	$0,0$	$1,0$	$2,0$	$3,0$
$-3,-1$	$-2,-1$	$-1,-1$	$0,-1$	$1,-1$	$2,-1$	$3,-1$
$-3,-2$	$-2,-2$	$-1,-2$	$0,-2$	$1,-2$	$2,-2$	$3,-2$
$-3,-3$	$-2,-3$	$-1,-3$	$0,-3$	$1,-3$	$2,-3$	$3,-3$

An arrangement for these number pairs is shown in Figure 5. Again, each dot in the array represents a pair of numbers. In order to indicate the number pair represented by each dot, we draw a broken line connecting the dots in the zero row and one connecting the dots in the zero column and then label the dots on the broken lines. The horizontal numerals name the first number represented by the dots, and the vertical numerals name the second number.

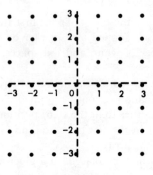

Figure 5

We could expand the array in Figure 5 to have it represent more number pairs. If we would continue this process indefinitely and think of representing all possible number pairs that exist, we would get a solid plane of dots as indicated in Figure 6. This plane is called the *number plane*. We think of this plane extending indefinitely in two dimensions. Therefore, all pairs of numbers you can think of are represented by the number plane. Of course, Figure 6 shows only a part of the number plane.

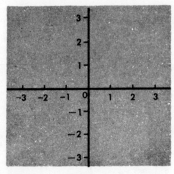

Figure 6

EXERCISE SET 1
Number Pairs

1. Write all the possible ordered pairs that can be formed from the numbers 1, 2, 3, 4. Draw an array that pictures these ordered pairs.
2. Draw an array like the one in Exercise 1 for each part of this

exercise and circle the dots in the arrangement that represent the number pairs that meet the conditions described in each part.

    a. The first member of the ordered pair is one more than the second.

    b. The first member of the ordered pair is twice the second.

    c. The second member of the ordered pair is greater than the first.

  3. Write all the possible ordered pairs that can be formed from the following numbers: -4, -2, 0, 2, 4. Show an array that pictures these ordered pairs.

  4. Do Exercise 2 using the arrangement described in Exercise **3**.

Locating Positions

    Suppose you live in a city where the streets are equally spaced and run either east and west or north and south. Suppose, also, that the blocks are all equal in length and that there are 8 blocks in 1 mile. Suppose further that the center of town is located at the intersection of Broadway and Main. This situation can be described by a scale drawing as shown in Figure 7.

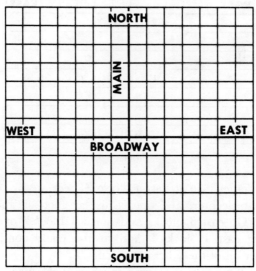

Figure 7

    What intersections are one-half mile from the center of town if you travel the one-half mile only on the streets — no cutting across lots, or going through alleys, or going out of your way? It isn't hard to find the answer. Since 1 mile equals 8 blocks, then one-half mile equals 4 blocks, and all you need to do is to find intersections on the map that are 4 blocks from the center of town. There are only 16 intersections that meet this condition, and they are lettered from *A* to *P* in Figure 8.

These 16 intersections that are marked by letters form a set of points. Because they are the only points on the map that satisfy the conditions given in the problem, they are called the *solution set* for the problem.

The next important question to consider is how you could give directions to anyone to go from the center of town to one of these intersections labeled by a letter. The most direct way to get to

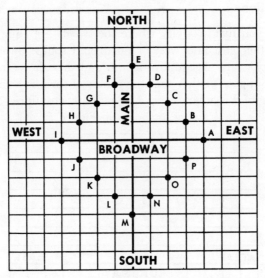

Figure 8

A from the center is to go east 4 blocks. There are two direct ways to *B* from the center, and one is just as good as another. You could say, "Go east 3 blocks and north 1 block," or "Go north 1 block and east 3 blocks." And so you could continue giving directions for all of the other intersections.

Let's see if we can use some mathematical ideas to simplify this matter of giving directions. The use of words such as *east*, *west*, *north*, and *south* is cumbersome and unnecessary. The signs + and − are good substitutes if everyone understands that + means east of Main or north of Broadway and − means the opposite: west of Main or south of Broadway. The next step might be to drop the names *Main* and *Broadway* and refer to these important landmarks as the *horizontal* and *vertical* axes. But *horizontal* and *vertical* are long words to write every time. Instead, we can make another simplification and call the horizontal axis the *x axis* and the vertical axis the *y axis*.

The center of the city, at the intersection of Main and Broadway, is the starting point for measuring blocks north or south and east or west. This intersection of the *x* and *y* axes is the *origin*.

Finally, we can abbreviate directions for locating points on the map. As you remember, when directions were given to locate *B*, there were two simple ones: "Go east 3 blocks and north 1 block," and "Go north 1 block and east 3 blocks." Rather than have two different direct ways to locate points, let's agree on one: When we write a location of a point as (3,1), we will know that the first number named tells how far to go horizontally and the second number named tells how far to go vertically. These numbers are called the *coordinates* of the point, the *x coordinate* always being the first, and the *y coordinate* being the second. Since a specific order is used to indicate the coordinates, they are called *ordered pairs*.

This system for naming points in a plane is illustrated in Figure 9. It is called a *rectangular coordinate system*. The plane, of course, is the number plane, sometimes called the *coordinate plane*.

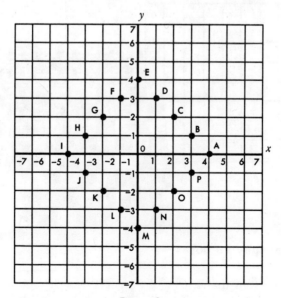

Figure 9

In the preceding section, we found that any number pair could be represented by a point in a number plane. Here we reversed the procedure and found that any point in a plane can be described by a pair of numbers.

The lettered points shown in Figure 9 can be described by writing the x and y coordinates as ordered pairs. For example, A can be described as $(4,0)$; B as $(3,1)$; E as $(0,4)$; F as $(-1,3)$; I as $(-4,0)$; J as $(-3,-1)$; L as $(-1,-3)$; M as $(0,-4)$; and P as $(3,-1)$.

EXERCISE SET 2
Naming and Plotting Points

1. Write the coordinates for each point at C, D, G, H, K, N, and O in Figure 9 as ordered pairs.

2. Draw a coordinate plane like the one in Figure 6 on page 6 and locate the following points in the plane.

a. $(1,1)$ d. $(3,-2)$
b. $(-2,1)$ e. $(2,-1)$
c. $(5,0)$ f. $(-3,-3)$

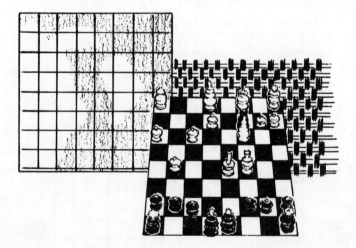

Graphs for Many Purposes

We have seen that a pair of numbers can be represented by a point in a plane and that a point in a plane can be described by a pair of numbers. By using a coordinate plane, it is possible to make drawings to represent number relationships and use number relationships to describe certain drawings. Drawings that illustrate the relationships between sets of numbers are called mathematical graphs.

The use of intersecting horizontal and vertical lines in such things as weaving, mapping, architecture, and the making of game boards for checkers and chess perhaps led man to the idea of using intersecting horizontal and vertical lines in a coordinate plane to draw pictures of numerical relationships. The graphing concept has many applications. One of the most important uses of graphs is to picture the relationship between sets of data. We are continually bombarded with collections of data that tell an important story about the world in which we live. In order to tell a story, however, the data needs to be organized. It might first be presented in tables as a set of number pairs. Sometimes, even more can be discovered by representing the data pictorially by means of a graph. Finally, it might be possible to find a simple mathematical equation that describes the data. Such a possibility is what a physical scientist hopes to do as he collects data to see if he can discover some new relationships.

You have worked with various collections of numerical data in your mathematics and science classes, and you have undoubtedly made use of many different types of graphs, such as bar graphs, circle graphs, and line graphs. But perhaps you have used graphs without fully understanding their use or meaning. You have already seen some of the ideas that serve as a basis for graphing. In the rest of Part III, we shall examine some important uses of, and mathematical ideas related to, graphing. The relationship between graphs and certain mathematical equations will be an important part of this study. Graphing will become much more meaningful and useful to you if you will gain an understanding of its mathematical nature.

On occasion, we have used the word *set* when discussing ideas related to graphs. The word *set* probably makes you think of a collection of objects, such as a set of books or a set of dishes. *Set* has the same meaning in mathematics, although in mathematics we are usually concerned with the collections of points or collections of numbers. Mathematicians often speak about the "set of all real numbers." This set includes all positive and negative whole numbers, fractions, irrationals (that is, numbers that cannot be expressed as a quotient of two whole numbers, such as $\sqrt{2}$), and *transcendental* numbers, such as π. We have spoken of a *solution set* of points, meaning all the points that satisfy a certain condition. When speaking of a set of numbers or points, it is important that information be given that will enable you to determine the numbers or points that belong to the set. We use braces $\{\ \}$ to indicate members of a set. For example, the set of whole numbers 1 through 4 would be written $\{1, 2, 3, 4\}$.

In the map problem of the preceding section, we worked with sets of ordered pairs of numbers. Let's consider a few more problems that might involve sets of ordered number pairs.

Suppose you have the problem of collecting and keeping a record of money from the sale of tickets for the class play. If the tickets are $2.00 each, you could make a table, as illustrated in Table 3, to compare the number of tickets sold and the amount of money that should be collected.

Table 3

Tickets	(x)	0	1	2	3	4	5	6
Money	(y)	0	$2	$4	$6	$8	$10	$12

If we assume that no student will sell more than 6 tickets, the table lists the set of ordered pairs that could result from our problem. This is the solution set for the problem. If we use the letter x as a placeholder for some number of tickets and the letter y as a placeholder for the correct number of dollars corresponding to the number of tickets, we can write a mathematical sentence, $y = 2x$, to describe the set of ordered pairs in this problem. As you know, such a sentence is often called an *equation* or a *formula*. It is important to remember that, in this problem, x can be replaced only by the number 0 or by the whole numbers 1 through 6. We call this set of numbers the *replacement set* for x. Any symbol, such as x or y, that holds a place for some number from a certain set of numbers is called a *variable*.

If we wish, we can use a rectangular coordinate system to picture the ordered pairs described by the table and the equation. This graph is shown in Figure 10. The coordinates of the points in the graph are the ordered pairs that meet the conditions of the problem.

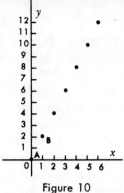

Figure 10

Note that the graph consists of a sequence of unconnected points. Point A represents the ordered pair $(0,0)$, and point B the ordered pair $(1,2)$. There are no points between A and B, for it is impossible to sell a fraction of a ticket. There are no points below A, for we have no concept of a negative ticket.

For another example, let's go back to the map problem of the previous section. Suppose you took away the restriction that you had to travel on city streets to find all points on a map that were 4 blocks from the center of town. Imagine yourself in flat, open country, and suppose you want to find all points on the ground that are exactly 4 blocks or 4 units from where you are standing.

Even though you have no streets, you can think of them as guide lines on a map, so the land could be pictured as a coordinate plane, as shown in Figure 11.

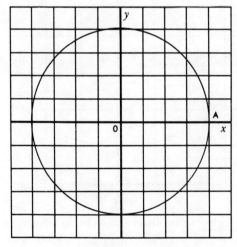

Figure 11

All points that are 4 units from point O will lie on the circumference of a circle of radius OA, which is 4 units long. This is a different solution set from that of the original problem because the restrictions have been changed. In the case in which you traveled only on city streets, you found just 16 points in the solution set. Now, with the restriction about following streets removed, there are more points than you can count in the solution set. We say that this solution set contains an *infinite* number of points. Since each point represents an ordered number pair, we can find an infinite number of ordered number pairs that will result from conditions of this problem. The question now arises, can we describe this situation with an algebraic equation? The answer is "yes," and it is easy to see how by working with Figure 12.

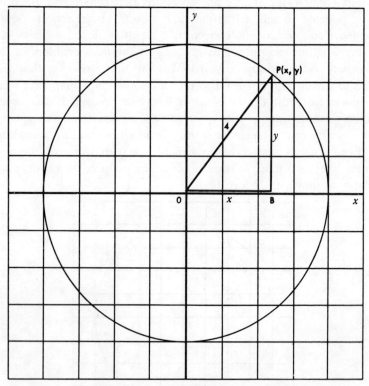

Figure 12

Draw the circle so that its center is at the origin O — the intersection of the x and y axes. Choose as point P any point on the circle. Draw the vertical line through P, intersecting the x axis at B. By definition, the coordinates of point P are the lengths of OB and BP. Let $OB = x$ and $BP = y$. The coordinates of point P, written as an ordered pair, are (x,y).

No matter what values x and y may have, they are restricted by the fact that OP must always be 4 units. But triangle OBP is a right triangle, and this suggests that the Pythagorean theorem can be used to express this restriction in algebraic form. As you know, the Pythagorean theorem states that the square of the hypotenuse of a right triangle is equal to the sum of the squares of the legs. The equation obtained from this theorem is

$$x^2 + y^2 = 4^2.$$

Therefore, any point P must have its coordinates (x,y) satisfy the equation $x^2 + y^2 = 4^2$ in order to be in the solution set for this problem. From the equation and the graph, we note another restriction. The replacement sets for x and y cannot contain numbers greater than 4 or less than -4.

We can use the equation to determine whether or not certain points lie on the circle.

Does the point having coordinates $(0,4)$ lie on the circle? Yes, because $0^2 + 4^2 = 4^2$ is a true statement.

Does the point having coordinates $(1,2)$ lie on the circle? No, because $1^2 + 2^2$ is not equal to 4^2. By locating the point $(1,2)$, you can see that it lies inside the circle.

Let's look at another number-pair situation. Suppose that a mathematical puzzle contains the condition, "One number is 2 more than another number." This could be written algebraically as

$$y = x + 2.$$

If the replacement set for x and y is the set of all real numbers, we could find an infinite set of ordered pairs that would meet the given condition. Table 4 lists a few of these ordered pairs that are in the solution set of $y = x + 2$. By representing enough of these ordered pairs with points in a coordinate plane, it becomes evident that the graph is a straight line, as shown in Figure 13 on page 142.

Table 4

x	-3	-2	0	1	$2\frac{1}{2}$	4
y	-1	0	2	3	$4\frac{1}{2}$	6

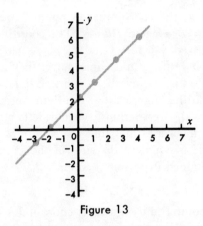

Figure 13

So far, you have seen graphs that had just a finite set of points in the solution set (like street intersections) or others that were smooth and unbroken like a circle or straight line. The next example is interesting because there are an infinite number of points in the solution set, but they are not smoothly connected.

A certain freight line charges $1.00 to deliver any package weighing less than 1 pound, $2.00 for any package weighing at least 1 pound but less than 2, $3.00 for any package weighing at least 2 pounds but less than 3, and so on. This statement also describes an infinite set of ordered pairs. The graph that pictures this infinite solution set is shown in Figure 14. A dot is placed at the left end of each line segment to indicate that the left end point of each segment (but not the right end point) is on the graph.

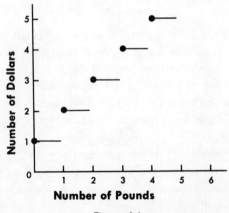

Figure 14

The line segments have not been connected, for this would indicate that the charge for a 1-pound package could be any value between $1.00 and $2.00, but, by definition, the charge is $2.00.

EXERCISE SET 3
Drawing Graphs

1. A certain number (y) is twice as large as another number (x).
   a. If the replacement set for x is the set of all real numbers, make a table listing five ordered pairs that meet the conditions of this problem.
   b. Write an equation that describes the set of ordered pairs under consideration.
   c. Draw a graph that pictures the set of ordered pairs.

2. In a grade-school athletic event, the distance recorded for participants in the standing broad jump was compared to the age of the participants. The results are summarized in the following table:

Age (yr.)	7	8	7	9	10	10	7	11	10	8	9
Distance (ft.)	2	3	$3\frac{1}{2}$	3	4	3	$2\frac{1}{2}$	5	5	$3\frac{1}{2}$	4

   a. Use a rectangular coordinate system to picture the ordered pairs given in the table. Represent ages on the x axis and distance on the y axis.
   b. Can you write an equation that describes this set of ordered pairs?

3. a. Make a table listing the ordered pairs pictured in the graph at the right.
   b. Write an equation (using the variables x and y) that describes the set of ordered pairs.
   c. List the members of the replacement set for x:

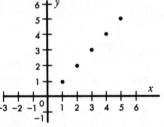

4. Which of the following points lie on the circle described by the equation $x^2 + y^2 = 4^2$? Which points lie inside the circle? Which points lie outside the circle?

   a. (0,0)   d. (2, − 4)   g. (− 3,2)
   b. (2,1)   e. (3, − 3)   h. (4,4)
   c. (− 1,1)   f. (0, − 4)

5. Write the equation that says all points (x, y) are 5 units from the origin of a coordinate plane; 6 units from the origin; 7 units from the origin; and, in general, r units from the origin.

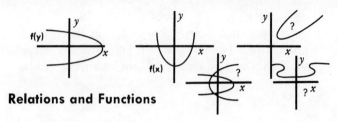

Relations and Functions

Let's examine the problems in the preceding section and in Exercise Set 3 and note some of the important ideas illustrated.

In each case, a set of ordered number pairs was obtained.

In each case, the set of ordered pairs was described by a verbal statement or by an algebraic equation, or both, and was pictured by a graph made up of points in a plane.

The graphs obtained were of various types. We found that a graph might consist of a series of unconnected points, a straight line, a series of unconnected line segments, or a closed curved line such as a circle. In all cases, the type of graph was determined by restrictions placed on the number pairs.

In mathematics, science, and business, it is often necessary to work with ordered pairs of numbers. It is important that we develop some appropriate methods and ideas for handling sets of ordered number pairs.

In mathematics, any *set of ordered number pairs* is called a mathematical *relation*. All of the sets of number pairs in the examples of the preceding section were mathematical relations. A mathematical relation in which knowledge of the first member of the ordered pair enables you to determine the second member is called a *function*.

The ordered pairs $\{ (1,2), (1,3), (1,4), (2,5), (2,6), (2,7) \}$ make up a mathematical relation, but not a function. Given the first member of one of the ordered pairs, say 1, you could not immediately determine the second member, for it might be 2, 3, or 4.

The set of ordered pairs $\{ (1,2), (2,3), (3,4), (4,5), (5,6), (6,7), (7,8) \}$ is both a relation and a function. If you are given a first member of an ordered pair, say 3, you can immediately determine 4 as the second member.

We have worked with four different ways of representing relations and functions:

1. We have used words to describe (or define) a set of ordered pairs.

Example: There are pairs of two real numbers such that the second number is 2 more than the first.

2. We have used algebraic symbols to describe the set of ordered pairs. The set of ordered pairs described by an equation having two variables is the set that will make the equation true, and is the solution set for the equation. *A solution set for a two-variable equation is a relation.* We must also remember that the solution set is also controlled by the replacement sets for the variables.

Example: There are ordered pairs such that $y = x + 2$, where the replacement set for x and $y = \{$all real numbers$\}$.

The function described by the equation $y = x + 2$ can be expressed as $\{(x,y) | y = x + 2)\}$. This is read, "The set of ordered pairs (x,y) such that y equals x plus 2." The vertical bar means "such that."

3. We have listed the ordered pairs in a table.

Table 5

x	1	$1\frac{1}{2}$	2	3	4	5
y	3	$3\frac{1}{2}$	4	5	6	7

Example: Table 5 is a partial list of ordered pairs described in the above examples.

4. We have used graphs to picture the set of ordered pairs.

Example: Figure 15 is a graph of the ordered pairs described in the above examples.

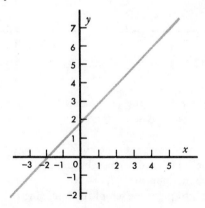

Figure 15

Hence we bring together ideas about ordered pairs, equations, and graphs.

EXERCISE SET 4
Working with Relations and Functions

1. Describe the sets of ordered pairs in questions 1, 2, 3, and 5 of Exercise Set 3 as relations or functions.

2. Which of the following sets of ordered pairs are functions?
    a. $\{(1,1), (2,2), (3,3), (4,4)\}$
    b. $\{(1,1), (1,2), (1,3), (1,4)\}$
    c. $\{(1,1), (2,1), (3,1), (4,1)\}$

3. Which of the following graphs picture functions?

a.
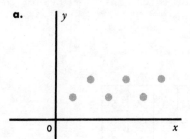

b.
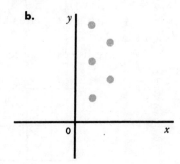

c.
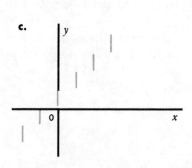

d.
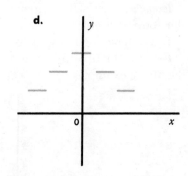

4. All three of the following statements tell you to draw the same graph. Draw that graph. Assume the replacement set = $\{$all real numbers$\}$

    a. Graph the solution set of the equation $y = 2x + 1$.
    b. Graph the function described by the equation $y = 2x + 1$.
    c. Graph the function $\{(x, y)|y = 2x + 1\}$.

5. Draw a graph to picture each of the following functions.

    a. $\{(x, y)|y = x\}$ Replacement set for x and $y = \{1, 2, 3\}$.
    b. $\{(x, y)|y = x + 1\}$ Replacement set for x and $y = \{-1, 0, 1\}$.
    c. $\{(x, y)|y = 3x\}$ Replacement set for x and $y = \{$all real numbers$\}$.

The Romance of Graphs and Equations

Descartes and Analytic Geometry

We have already noted some interesting comparisons between equations and graphs. The principles that we have observed are not new, however. René Descartes (1596-1650), a French mathematician, discovered that a geometric picture that corresponds to an algebraic equation can be drawn. November 10, 1619, is the official birthday of this idea and might also be celebrated as the birthday of modern mathematics. Without this new idea, progress in science and engineering could never have taken place the way it has. Descartes opened up a new direction for mathematics. Algebra and geometry were linked together for the first time by a simple idea — a rectangular coordinate system, known also as *Cartesian coordinates*, in honor of Descartes. Analytic geometry became the name of this new branch of mathematics, and it soon became a close partner with all branches of mathematics.

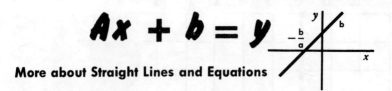

More about Straight Lines and Equations

Figure 16 illustrates a steep hill that rises at a uniform rate. Suppose you want to go from position P_1 at the bottom of the hill to a position P_2 on the hill. As you move up the hill to P_2, you move a certain distance vertically and a certain distance horizontally. Line segment P_2D represents the vertical distance you would move, and P_1D represents the horizontal distance. Since the hill rises at a uniform rate, these two values can be used as a measure of the *steepness* of the hill. The hill *rises* vertically a distance of P_2D for a horizontal distance P_1D. The ratio $\dfrac{P_2D}{P_1D}$ gives a measure called the *slope* of the hill. The slope expresses the steepness of the hill. *The greater the slope, the steeper the hill.*

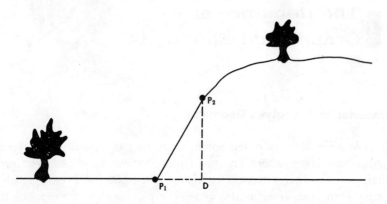

Figure 16

Let's translate this situation to a coordinate plane as shown in Figure 17. Suppose P_1 is a point having coordinates (1,2) and P_2 is a point having coordinates (3,6). Joining the points P_1 and P_2 forms a line which makes a definite slope with respect to the x axis. The slope can be measured by finding the ratio of the vertical change to the horizontal, just as it was done for the hillside. Each of these changes can be expressed in terms of the co-ordinates of the points.

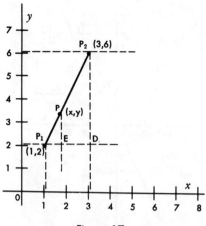

Figure 17

In the figure, $P_2D = 6 - 2 = 4$ (the vertical change) and $P_1D = 3 - 1 = 2$ (the horizontal change).

The ratio of these two changes, the vertical to the horizontal, is $\frac{4}{2}$ or $\frac{2}{1}$; that is, there are 2 units of vertical change upward for every unit of horizontal change to the right.

Now you might ask how you can find points P in the plane, such that the slope of P_1P is the same as P_1P_2. Since P, theoretically, could be anywhere in the plane, you can call its coordinates (x,y). The vertical change EP can be written as $(y - 2)$. The horizontal change P_1E can be written as $(x - 1)$, and the ratio is $\frac{y - 2}{x - 1}$.

We recall that this ratio is equal to $\frac{P_2D}{P_1D}$ or $\frac{2}{1}$. Therefore, the algebraic condition of equal slope provides this statement:

$$\frac{y - 2}{x - 1} = \frac{2}{1}.$$

This can be simplified to $\quad y - 2 = 2(x - 1)$.

Solving for y, we have $\quad y = 2x - 2 + 2$,

or $\quad y = 2x$.

This equation describes the coordinates of the set of points that make up the line shown in Figure 17. For the sake of convenience, we just say that it is the equation of the line. Figure 18 shows a more complete picture of this line. We can check the accuracy of this drawing by assigning values to x and then using the equation $y = 2x$ to compute corresponding values for y.

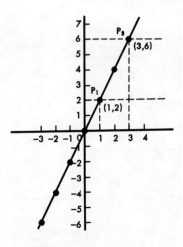

Figure 18

Consider now a more general situation, in which we have a line joining two points $P_1(x_1, y_1)$ and $P_2(x_2, y_2)$, with $P(x, y)$ as any other point on the line. As we see in Figure 19, the ratio of the

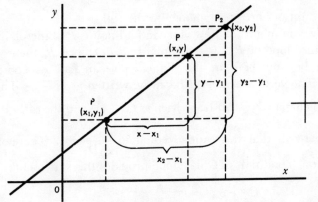

Figure 19

vertical change to the horizontal change can be written as $\dfrac{y_2 - y_1}{x_2 - x_1}$ or $\dfrac{y - y_1}{x - x_1}$. Since these two ratios are equal, we have

$$\frac{y - y_1}{x - x_1} = \frac{y_2 - y_1}{x_2 - x_1}.$$

Therefore, by knowing the coordinates of two points on any straight-line graph that pictures a function, we can determine an equation that describes the graph and, thus, define the function.

For example, suppose a straight line passes through the points (1,2) and (2,5). Let $x_1 = 1$, $y_1 = 2$, $x_2 = 2$, and $y_2 = 5$. Using the above formula we have

$$\frac{y - 2}{x - 1} = \frac{5 - 2}{2 - 1},$$
$$\text{or } y - 2 = 3(x - 1),$$
$$\text{or } y = 3x - 1$$

as the equation of the line.

The ratio of the vertical change to the horizontal change (the *slope*) of the line is often represented by the symbol *m*. From Figure 17,

$$\frac{y - y_1}{x - x_1} = m = \text{the slope,}$$
$$\text{or } y - y_1 = m(x - x_1).$$

If we know the slope and the coordinates of one point of a straight-line graph, we can write an equation that describes the graph.

For example, suppose a line passes through the point $(2, -1)$ and has a slope of 3. Let $x_1 = 2$, $y_1 = -1$, and $m = 3$. Then we have

$$y - (-1) = 3(x - 2),$$
$$\text{or } y + 1 = 3x - 6,$$
$$\text{or } y = 3x - 7$$

as the equation of the line.

Let's take another example. A straight-line graph is pictured in Figure 20. We see that the line passes through (0,2) and that for every change of positive 3 units horizontally there is a change of positive 2 units vertically. Therefore, $x_1 = 0$, $y_1 = 2$, $m = \dfrac{2}{3}$, and we have

$$y - 2 = \frac{2}{3}(x - 0),$$

$$\text{or } y - 2 = \frac{2}{3}x,$$

$$\text{or } y = \frac{2}{3}x + 2.$$

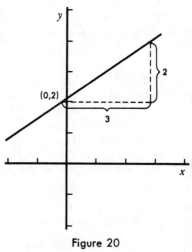

Figure 20

In this example, the point selected had the coordinates $(0,2)$. This point is on the y axis and indicates the point at which the line crosses the y axis. Such a point is called the y *intercept*.

In general. if we are given the slope of a line (m) and the co-ordinates of the y intercept $(0,b)$, the equation that describes the graph would be

$$y - b = m(x - 0),$$
$$\text{or } y - b = mx,$$
$$\text{or } \quad y = mx + b.$$

Therefore, if we know the slope, m, and the y intercept, b, of a straight-line graph, we can write the equation for the graph immediately as: $y = mx + b$.

Look at the graph in Figure 21. We can see that the y intercept of the graph is 2, but what is the slope? We note that an increase of one unit horizontally will result in a corresponding *decrease* of two units vertically. The ratio of the vertical change to horizontal change is $\dfrac{-2}{1}$. The slope of the line is -2. The equation for the line is $y = -2x + 2$.

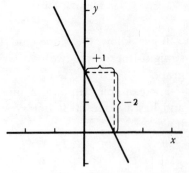

Figure 21

Notice that it was possible to write the equations in all the examples in the form $y = mx + b$.

EXERCISE SET 5
Writing Equations

1. Find the equations of the lines joining the following pairs of points.
a. $(2,1)$ and $(3,4)$ b. $(-1,2)$ and $(0,-1)$
c. $(-2,5)$ and $(-1,5)$ d. $(a,0)$ and $(0,b)$

2. Find the equations of the lines having the following slopes and y intercepts.
a. slope = 1, y intercept = 1
b. slope = $\dfrac{1}{2}$, y intercept = 5
c. slope = -2, y intercept = -3

3. Find the equations of the lines that pass through the given points and have the given slopes. Write the equations in the form $y = mx + b$.

  a. $(-1,2)$, slope $= 2$ b. $(0,5)$, slope $= 3$

  c. $(0,-2)$, slope $= \dfrac{-2}{3}$ d. $(a,0)$, slope $= m$

4. Find the equations that describe the graphs shown below.

a.

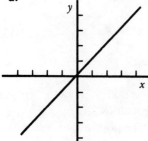

d.

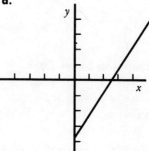

b.

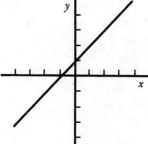

e.

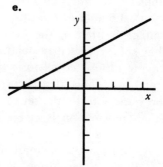

c.

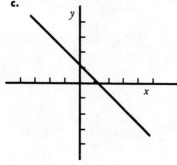

f.

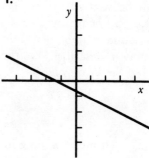

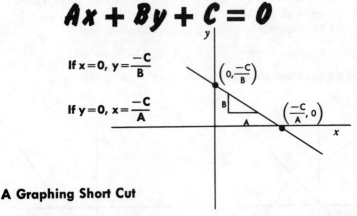

$$Ax + By + C = 0$$

If $x = 0$, $y = \dfrac{-C}{B}$

If $y = 0$, $x = \dfrac{-C}{A}$

$\left(0, \dfrac{-C}{B}\right)$

$\left(\dfrac{-C}{A}, 0\right)$

A Graphing Short Cut

We're always looking for an easy way to do a job. Let's attempt to find an easy way to draw straight-line graphs.

We have observed that the equation corresponding to a straight-line graph can be written in the form $y = mx + b$, m being the slope of the line and b the y intercept. It seems sensible that we should be able to use this idea in reverse. That is, if an equation can be written in the form $y = mx + b$, perhaps we can assume that the graph corresponding to the equation is a straight line having a slope equal to m (the coefficient of x) and a y intercept equal to b. This idea is correct, if we assume the replacement set for x and y to be the set of all real numbers. From here on, if no other information is given, we shall assume the replacement set for variables to be the set of real numbers.

As you have seen, mathematical functions can be described by equations. The solution set of ordered pairs of an equation that can be written in the form $y = mx + b$ is called a *linear function*. Thus, a linear function is defined as $\{(x,y)\,|\,y = mx + b\}$.

Consider the linear function $\{(x,y)\,|\,2y - 3x = 6\}$; that is, the function described by the equation $2y - 3x = 6$. This equation could be written as

$$2y = 3x + 6,$$
$$\text{or} \quad y = \frac{3}{2}x + 3.$$

This equation is now in the form $y = mx + b$, where $m = \dfrac{3}{2}$ and $b = 3$. If we assume the replacement set for x to be the set of real numbers, the graph that pictures the solution set of $y = \dfrac{3}{2}x + 3$ will be a straight line having a slope of $\frac{3}{2}$ and a y intercept of 3. To draw the graph, we first locate the point (0,3), as shown in Figure 22. From that point, we move 2 units horizontally and then 3 units vertically, locating a second point (2,6). We then draw the line joining the two points. This is the *slope-intercept* method of graphing.

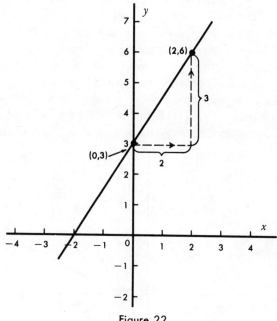

Figure 22

The straight line in Figure 22 is the graph of the solution set of $y = \frac{3}{2}x + 3$. For convenience, we merely call it the *graph of the equation*.

Let's bring together our ideas about slope, equations, and lines by examining several lines going through the point (1,2) with different slopes and showing them on one rectangular coordinate system.

$y = \frac{1}{2}x + \frac{3}{2}$; the slope is $\frac{1}{2}$ $y = -\frac{1}{2}x + \frac{5}{2}$; the slope is $-\frac{1}{2}$

$y = x + 1$; the slope is 1 $y = -x + 3$; the slope is -1

$y = 2x$; the slope is 2 $y = -2x + 4$; the slope is -2

$y = 0x + 2$; the slope is 0

$0y = x - 1$; the slope is undefined

In this last equation, any replacement for y will give the same value, 1, for x. Therefore, there can be no horizontal change on the graph. Since slope is defined as the ratio of a vertical change to a corresponding horizontal change, the slope will be undefined, for division by 0 is undefined.

All of the lines described by the above equations have one point in common, but each one has a different slope. Since they have something in common, they are called a family of lines.

Using the graphs in Figure 23, you can compare lines of positive slope with those of negative slope. If you observe the lines as you move horizontally from left to right, you can see that a positive slope shows y values increasing as x values increase, and a negative slope shows y values decreasing as x values increase. Zero slope shows no change in y. Trace the path shown by the arrows in Figure 23, and notice the way the slopes of the lines change.

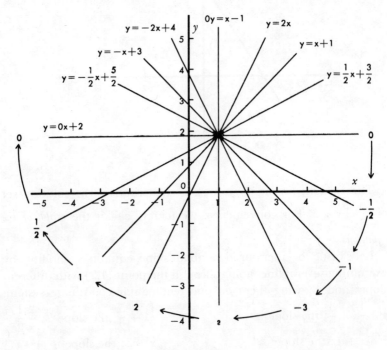

Figure 23

EXERCISE SET 6
Graphing Linear Functions

1. Find the slope and the y intercept of the graph that corresponds to each of the following equations.

a. $y = 2x - 3$ b. $2y = 3x + 4$

c. $x + y = 4$ d. $2x - 3y = 5$

2. Use the slope-intercept method to graph the linear functions described by the following equations.

a. $y = x - 1$ b. $2x + y = 5$ c. $0y = x - \dfrac{1}{2}$

3. Graph the solution sets for the following equations.

a. $y = \dfrac{1}{2}x - 3$ b. $3y - 6 = -x$ c. $4y = 0x + 8$

4. Observations have shown that the Fahrenheit temperature F equals 39 added to the number of chirps C of the common black cricket in 15 seconds. Write the equation that fits this statement. Make a table of values that are reasonable, remembering that warm summer evenings will be the best time to verify such data. Then graph the data that has been verified.

5. Harlow Shapley, an eminent astronomer, made a study of the ant which has the technical name of *Tapinoma sissila*. He found that the ant's speed S in inches per minute and the Fahrenheit temperature F are related by the formula

$$F = 11S + 39.$$

Make a table of values from this formula. Graph the data.

6. Using the information from Problems 4 and 5, see if you can answer the question: If a cricket chirps 100 times a minute, how fast does the ant run?

$$C^2 = A^2 + B^2$$

$$C = \sqrt{A^2 + B^2}$$

Distance Between Points on a Graph

The same horizontal and vertical changes that were used to find the slope of the line joining 2 points in a Cartesian plane can also be used to find the distance between the 2 points. This is because these changes can be represented geometrically as legs of a right triangle and the Pythagorean theorem can then be used to find the hypotenuse. Suppose the two points are the ordered pairs (1,2) and (3,6) as shown in Figure 24. The vertical change is $6 - 2 = 4$. The horizontal change is $3 - 1 = 2$.

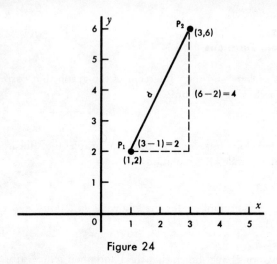

Figure 24

Instead of using these values in a ratio to find the slope of the line, you use them as legs of a right triangle so that

$$d^2 = 2^2 + 4^2 \quad \text{or} \quad d^2 = 20$$
$$d = \sqrt{20}, \text{ or } 4.47 \text{ units.}$$

In general, for any 2 points $P_1(x_1,y_1)$ and $P_2(x_2,y_2)$, the legs of the right triangle will be $(x_2 - x_1)$ and $(y_2 - y_1)$, and the distance between them can be written as

$$d^2 = (x_2 - x_1)^2 + (y_2 - y_1)^2,$$
$$\text{or } d = \sqrt{(x_2 - x_1)^2 + (y_2 - y_1)^2}.$$

Although this seems to be a very complex formula, it is easy to use. Suppose you are given the coordinates of two points, (1,2) and (3,5). To find the distance between the points, let $x_1 = 1$, $y_1 = 2$, $x_2 = 3$, and $y_2 = 5$. Then substituting in the formula, we have

$$d = \sqrt{(3 - 1)^2 + (5 - 2)^2} \quad \text{or} \quad d = \sqrt{13}, \text{ or } 3.6 \text{ units.}$$

EXERCISE SET 7
Distance Problems

1. Show that the points $P_1(0, 5)$, $P_2(6, -3)$, and $P_3(3, 6)$ are vertices of a right triangle. (Hint: Use the Pythagorean theorem.)

2. Show that the points $P_1(8, 9)$, $P_2(-6, 1)$, and $P_3(0, -5)$ form an isosceles triangle. (Hint: Show two sides are equal.)

3. Show that the quadrilateral whose vertices are $P_1(-7, 7)$, $P_2(2, 0)$, $P_3(10, 3)$, and $P_4(1, 10)$ is a parallelogram. (Hint: Show opposite sides are equal.)

The Case of the Buried Treasure

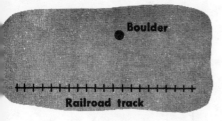

Boulder

Railroad track

Figure 25

A treasure map has been torn and the bottom half is missing, but you can still read this much: "The treasure is buried in a spot that is the same distance from the boulder that it is from the railroad track. It is also . . ." and the rest of the information is missing

Here is a frustrating situation. A fortune is at your finger tips, but you don't know how to find the treasure.

However, you might use the information that is given and see just how far this can take you toward the solution.

First of all, the distance of the treasure from the railroad track is interpreted as being the length of a perpendicular drawn to the tracks from the treasure. It seems logical, as a first step, to set up a coordinate system and try to use it to solve the problem. Perhaps a logical way to set up the coordinate plane is to let the track (assuming it is straight) be the *x* axis and let the boulder be on the *y* axis at point *B*, as shown in Figure 26. Suppose the boulder is 50 feet from the track, placing it at the point (0,50) in the coordinate plane. Now, you want to find some points that are the same distance from point *B* and the *x* axis. You can easily see that one such point is located at (0,25) in the coordinate plane.

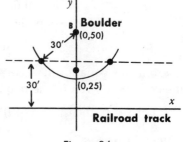

Figure 26

By marking a line parallel to the railroad track, at say, 30 feet from it, and marking a circle with a 30-foot radius with its center at *B*, you could find two more points that meet the conditions given in the map. In a similar fashion, you can find other points that will meet the conditions. Since each point in the coordinate plane can be represented by an ordered pair of numbers, let's see if an equation can be found to describe the set of ordered pairs and help solve the problem.

You know that the treasure is located at some point P having coordinates (x,y), but you do not know the exact location of this point. In Figure 27, P is placed at any convenient point in the coordinate plane. Now you know that PB, the distance from the treasure to the boulder, must equal PT, the perpendicular distance from the treasure to the railroad track. But PB is the distance between the points $B(0,50)$

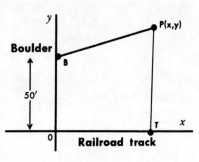

Figure 27

and $P(x,y)$. Using the formula for the distance between two points, you have

$$PB = \sqrt{(x - 0)^2 + (y - 50)^2}.$$

The distance PT is equal in value to the y coordinate of P; that is, $PT = y$.

Since $PT = PB$, the algebraic restriction expressing the solution set for all ordered pairs (x,y) is

$$\sqrt{(x - 0)^2 + (y - 50)^2} = y.$$

Squaring both sides and collecting like terms, you get
$$x^2 + y^2 - 100y + 2500 = y^2,$$
which simplifies to $100y = x^2 + 2500,$

or, finally, $y = \dfrac{1}{100}x^2 + 25.$

From this equation it is easy to find pairs of values that satisfy the condition given in the map.

Let $x = 0$, then $y = \dfrac{1}{100}(0)^2 + 25 = 25$ ft.

Let $x = 10$ ft., then $y = \dfrac{1}{100}(10)^2 + 25 = 1 + 25 = 26$ ft.

Let $x = 20$ ft., then $y = \dfrac{1}{100}(20)^2 + 25 = \dfrac{400}{100} + 25 = 29$ ft.

You can also let x have negative values.

Let $x = -10$ ft., then $y = \dfrac{1}{100}(-10)^2 + 25 = 1 + 25 = 26$ ft.

Let $x = -20$ ft., then $y = \dfrac{1}{100}(-20)^2 + 25 = \dfrac{400}{100} + 25 = 29$ ft.

The graph of this data is shown in Figure 28.

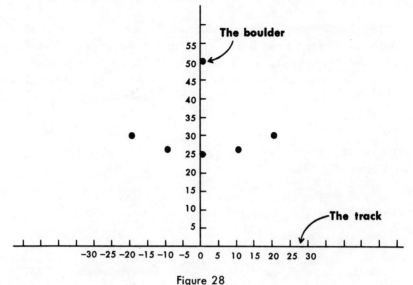

Figure 28

If you fill in more data, you can see that the solution set of points becomes a smooth curve. So there is an infinite number of points where that treasure might be. You had better give up looking for the treasure unless you find the other part of the instructions, or unless you want to dig a very long ditch like the one in Figure 29.

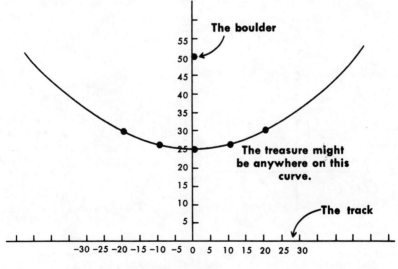

Figure 29

In general, points that are always the same distance from a fixed point and a fixed line lie on a kind of curve that is called a *parabola*. The solution set of any equation of the form $y = ax^2 + bx + c$ (where a cannot be 0) is a function that can be graphed as a parabola. Figure 29 pictures part of a parabola.

EXERCISE SET 8
Working with Parabolas

1. Complete this table of values for each of the following equations and then graph the complete solution set of the equation.

x	-3	-2	-1	0	1	2	3
y							

All of them will be parabolas like that on the treasure map.

a. $y = x^2$

b. $y = x^2 + 2$

c. $y = x^2 - 2$

d. $y = 2x^2$

e. $y = \frac{1}{2}x^2$

f. $y = x^2 + 2x$

g. $y = x^2 - 2x$

h. $y = x^2 + 2x + 1$

2. Write an equation that describes the coordinates of all these points:

a. The same distance from the x axis as from the point $(0,4)$.

b. The same distance from a line parallel to the x axis but two units below it as from the point $(1,2)$.

3. Given two fixed points $P_1(-3, 0)$ and $P_2(3, 0)$, find points P such that $P_1P + P_2P = 10$.

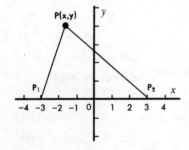

Suggestion: Place thumbtacks at P_1 and P_2. Tie string at P_1 and P_2 so that the total length of the string is 10 units long. Trace curve with pencil at P, being sure that you keep the string tight.

The curve that you have traced is called an *ellipse*.

4. Find out all you can about the following curves:

a. *hyperbola*   b. *cycloid*   c. *catenary*

$P(x, -y, z)$

Graphing in Space

As you well know, this is the space age, but you perhaps do not know that many of the ideas of mathematics can be applied to three-dimensional space.

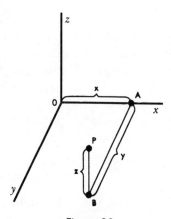
Figure 30

You have seen how a point can be located in a plane by giving its coordinates as an ordered pair. Let's see how points in space can be located by adding a third axis z at right angles to a pair of x and y axes as shown in Figure 30. A point P can be located by an ordered triple (x, y, z).

To get to P from the origin O, you can travel down the x axis for x units from O to A, and then parallel to the y axis from A to B for y units, and finally parallel to the z axis from B to P for z units. This is similar to the type of directions you might give someone in telling him how to locate an office in a skyscraper.

Negative coordinates can be set up just as they were in two dimensions. The ordered triple $(-2, 3, -1)$ can be found at point A in Figure 31. The ordered triple $(3, -2, 1)$ can be found at point B.

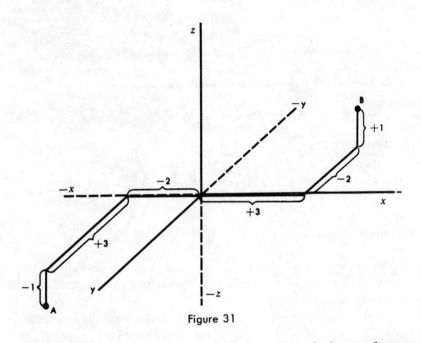

Figure 31

You discovered that, in two dimensions, the solution set for an equation such as $y = 3x - 2$ can be pictured as a straight-line graph if the replacement set for the variables is the set of real numbers. But what can be said about the three-dimensional graph of the solution set for an equation like $x + y + z = 4$ if the replacement set for the variables is the set of real numbers?

To answer this question, you will have to change this equation into a simpler form that you can handle. First, let's replace z with 0. You then have

$$x + y + 0 = 4,$$
or $x + y = 4$.

If you graph this equation in the xy plane, you will obtain the line c shown in Figure 32. This line will be part of the graph of $x + y + z = 4$.

Now let's replace x with 0, getting

$$y + z = 4.$$

If you graph this in the yz plane, you obtain line a in Figure 32, which will also be part of the graph of $x + y + z = 4$.

Finally, replacing y with 0, you get the equation

$$x + z = 4.$$

Graphing this in the xz plane gives line b of Figure 32 which is also a part of the graph of the equation $x + y + z = 4$.

These three lines determine three points (A, B, and C) in the xyz coordinate space, and these three points determine a plane. Therefore, the lines a, b, and c are in the same plane, and the graph of $x + y + z = 4$ is the entire plane determined by the points A, B, and C.

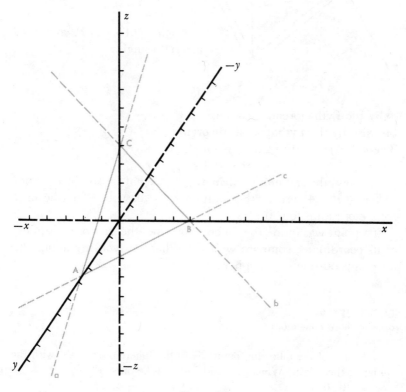

Figure 32

In other words, the plane passing through points A, B, and C is made up of all the points (and only those points) whose coordinates make up the solution set for the equation $x + y + z = 4$.

Also, you found a way in two dimensions to state the condition for all points in a plane 4 units from a fixed point. When the fixed point was the origin, the solution turned out to be all ordered pairs that satisfied the equation

$$x^2 + y^2 = 4^2.$$

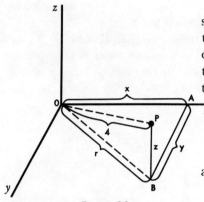
Figure 33

What do you have for a solution set for points in space that are to be four units from the origin? To answer this, you need to use the Pythagorean theorem twice. Figure 33 shows:

$$OA = x, \qquad OB = r,$$
$$AB = y, \qquad OP = 4,$$
$$BP = z.$$

Angles OAB, ABP, and OBP are right angles.

By the Pythagorean theorem, $x^2 + y^2 = r^2$, but also by the Pythagorean theorem, $r^2 + z^2 = 4^2$. Therefore, substituting $x^2 + y^2$ for r^2, you have

$$x^2 + y^2 + z^2 = 4^2.$$

This says that all number triples (x, y, z) that satisfy the equation $x^2 + y^2 + z^2 = 4^2$ are in the solution set. All such points lie on a sphere of radius four units.

If a point was inside the sphere, how does the sum of the squares of its coordinates compare with 4^2? What can you say about this for points outside of the sphere?

EXERCISE SET 9
Graphs in Three dimensions

1. Which of the following points lie *on* the sphere of radius 4 with the center at the origin? Which lie inside? Which lie outside?

a. $(0, 0, 0)$ e. $(1, -3, 2)$
b. $(2, 1, 1)$ f. $(0, 0, -4)$
c. $(0, 4, 0)$ g. $(1, 1, 1)$
d. $(-3, 3, 3)$ h. $(5, 0, 1)$

2. At what point does the graph for the equation $x + y - z = 3$ cut the x axis? The y axis? The z axis?

3. At what point does the plane graph for $2x - y = 6 - 3z$ cut the x axis? The y axis? The z axis?

4. Write an equation describing the coordinates of the points on a sphere of radius 2.5 with the center at the origin.

From Equations
to Inequalities

Picturing Inequalities

If you have ever ordered merchandise from a mail-order catalog, you have probably seen a map like the one in Figure 34. A map such as this is used to determine the postage to be paid on an order. For example, if you lived within a 100-mile radius of Kansas City, you would pay the postage rates for zone 1. In other words, anyone living 100 miles or *less* than 100 miles from Kansas City will pay the same shipping rates on orders. This example illustrates one of many uses of inequalities in everyday life.

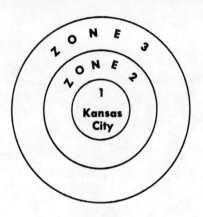

Figure 34

We will use the following mathematical symbols to take the place of words:

The symbol, $<$, is read "is less than."
The symbol, $\leq$, is read "is less than or equal to."
The symbol, $>$, is read "is greater than."
The symbol, $\geq$, is read "is greater than or equal to."
The symbol, $\neq$, is read "is not equal to."

The sentence, $5 < 8$ is read, "5 is less than 8," and $5 > 3$ is read "5 is greater than 3." The statement, $5 + 3 \neq 10$, is read "5 plus 3 is not equal to 10." These are all mathematical statements that use either the equality or the inequality symbols. In a statement of the type, $x < 8$, you would be looking for the set of replacements for x that would make the statement true. As you know, such a set is called the solution set. If the replacement set for x is restricted to the natural numbers, 1, 2, 3, . . ., then the solution set can be graphed on a number line as the heavy dots at 1, 2, 3, 4, 5, 6, and 7 in Figure 35.

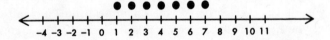

Figure 35

If the replacement set for x has no restrictions on the number line, then the solution set can be graphed as the heavy line to show all real numbers less than 8, as in Figure 36.

The open circle indicates that 8 is not included in the graph.

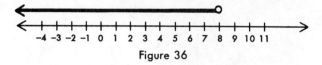

Figure 36

Now, suppose you want to work with the set of ordered pairs (x, y) governed only by the restriction $x < 8$, with no restriction on y. This set could be written as $\{(x, y) | x < 8\}$. The graph that pictures this set of ordered pairs is shown in Figure 37. The coordinates of all the points in the shaded part of the coordinate plane left of line AB would be members of the desired set. Here we have a graph that is a part of a plane.

Is $\{(x, y) | x < 8\}$ a mathematical relation? Is this set a function? The answer to the first question is, of course, "yes," for a relation is a set of ordered pairs. The answer to the second question is "no," for knowing a value of x does not enable us to determine a specific value for y.

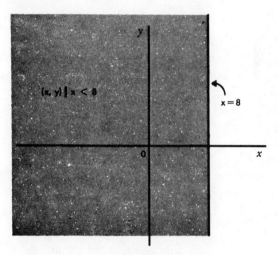

Figure 37

In a three-dimensional situation we might be concerned with a set such as $\{(x, y, z) | x < 8\}$; that is, the set of ordered triples (x, y, z) such that x is less than 8, with no restrictions on y or z. The graph of this set, shown in Figure 38, is made up of all points to the left of plane $ABCD$.

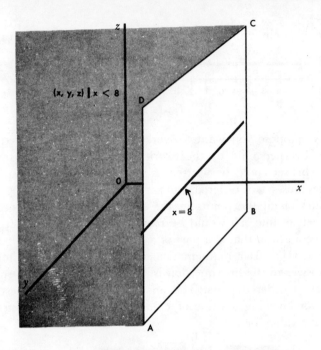

Figure 38

To be able to handle more complex expressions that involve inequalities, it is necessary to know that the usual assumptions used in working with equations can also be used with inequalities. One exception is that of multiplying or dividing both sides of an inequality by the same negative quantity. In such cases, the inequality is reversed; that is, a $>$ symbol would be replaced by a $<$ symbol and vice versa.

Let's demonstrate the assumptions that apply to inequalities with a simple example. If $4 < 9$ is true, the following are true:

$4 + 2 < 9 + 2$	Adding the same quantity to both sides *does not* reverse the inequality.
$4 - 2 < 9 - 2$	Subtracting the same quantity from both sides *does not* reverse the inequality.
$2 \times 4 < 2 \times 9$	Multiplying by the same positive quantity on both sides *does not* reverse the inequality.
$- 2 \times 4 > - 2 \times 9$	Multiplying each side by the same negative quantity *does* reverse the inequality.

$$\frac{4}{2} < \frac{9}{2}$$

Dividing each side by the same positive quantity *does not* reverse the inequality.

$$\frac{4}{-2} > \frac{9}{-2}$$

Dividing each side by the same negative quantity *does* reverse the inequality.

These assumptions can be used to "solve" an inequality. For example, we can algebraically find the replacements for x that will make the following inequality true:

$$\frac{2x}{-3} - 1 > 5$$

$$\frac{2x}{-3} > 6$$

Adding the same quantity does not reverse the inequality.

$$2x < -18$$

Multiplying by the same negative quantity reverses the inequality.

$$x < -9$$

Dividing by the same positive quantity does not reverse inequality.

Let's look now at the relation described by $y < x$. If the replacements for x and y must be selected from the natural numbers 1 through 4, then $\{(x, y) \,|\, y < x\} = \{(2,1), (3,1), (3,2), (4,1), (4,2), (4,3)\}$. The graph that pictures this solution set is shown in Figure 39.

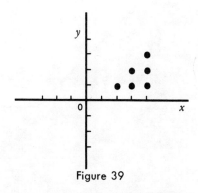

Figure 39

If the replacement set for the variables is the set of real numbers, the graph of the solution set of $y < x$ can be found by drawing the straight-line graph of $y = x$. Then, every point *below* the line has a y coordinate with a value less than the value of its x coordinate. The shaded part of Figure 40 shows the solution set of $y < x$.

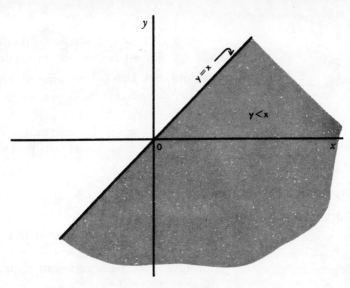

Figure 40

The solution set of an inequality such as $x + y < 4$ is handled by solving for y:

$$y < 4 - x.$$

The solution set is shaded below the line $y = 4 - x$, as shown in Figure 41.

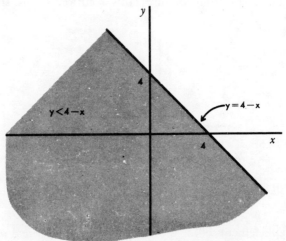

Figure 41

In the case $x^2 + y^2 < 4^2$, the solution set is the inside of the circle $x^2 + y^2 = 4^2$, as shown in Figure 42.

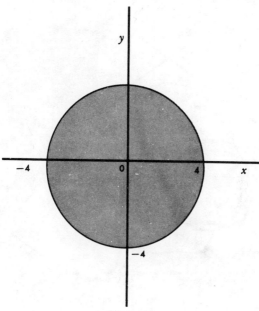

Figure 42

If the boundary is to be included as well as the shaded portion inside, the expression is written as

$$x^2 + y^2 \leq 4^2.$$

EXERCISE SET 10
Working with Inequalities

1. Solve the following inequalities algebraically, and then graph each solution set on a number line.

a. $x - 3 > 7$

b. $3x + 5 < 2$

c. $\dfrac{x}{4} - 1 < 15$

d. $-2x > 8$

e. $\dfrac{x}{-3} - 5 < -7$

f. $2x - 3 < 3x + 2$

2. Draw the graph of the mathematical relation described by the following inequality if the replacement set for the variables is the set of natural numbers 1 through 5.

$$y > x + 1$$

3. Draw the graph of the solution set for each of the following inequalities. Assume the replacement set to be the set of real numbers.

a. $y > x$

b. $y > x + 7$

c. $y < x^2$

d. $x^2 + y^2 > 4^2$

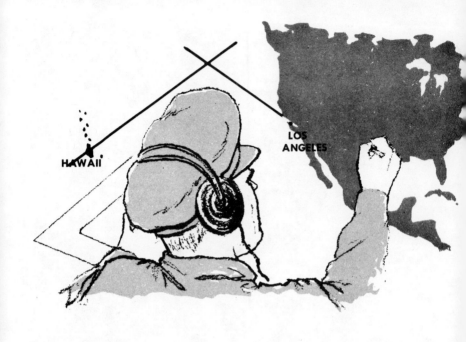

The Mathematics of Intersecting Graphs

"A number is 2 more than a second number and is also 6 less than twice the second number. What is the number?"

There are a number of methods that can be used to solve this simple number puzzle, but let's turn our attention to a graphing method. There are two conditions given in this problem. If we let y hold a place for the first number and x hold a place for the second, we can express the two conditions with the equations

$$y = x + 2 \text{ and } y = 2x - 6.$$

The graphs of the solution sets for the two equations are shown in Figure 43. The two graphs intersect at point P. The coordinates of P make up an ordered pair of numbers that belong to the solution sets of both equations. Therefore, the ordered pair $(8,10)$ meets both conditions of the problem, and the answer to the problem is 10. The graph enabled us to find a common member of the two solution sets. A set of numbers obtained from the common members of two solution sets is called an *intersection set*.

The symbol, $\cap$, is used to indicate an intersection of two sets. Thus, $\{(x, y)|y = x + 2\} \cap \{(x, y)|y = 2x - 6\}$ indicates the intersection of the solution sets of $y = x + 2$ and $y = 2x - 6$.

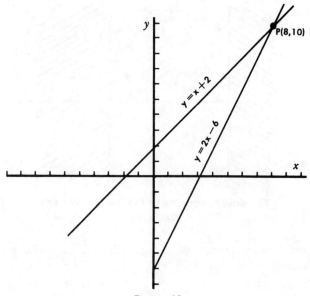

Figure 43

Let's consider more examples in which we will want to find numbers or a set of numbers that meet two mathematical conditions at the same time. We can impose conditions of inequality, also.

If we are working with just one variable, and if the two conditions imposed are

$$x < 4$$
$$\text{and} \quad x > -4,$$

then the two graphs have points in common, as shown in Figure 44. The solid line is the graph of the solution set for $x < 4$. The dotted line is the graph of the solution set for $x > -4$.

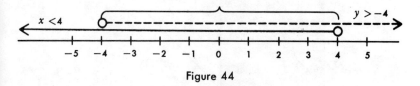

Figure 44

The bracketed portion is the graph of the intersection set because it is common to both.

In two dimensions, the graph of $\{(x, y) | x < 4\} \cap \{(x, y) | x > -4\}$ is a strip between $x = 4$ and $x = -4$, as shown in Figure 45.

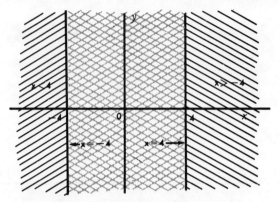

The double shading is the intersection set.
Figure 45

In three dimensions, the graph of the intersection set of $\{(x, y, z) \mid x < 4\}$ and $\{(x, y, z) \mid x > -4\}$ would be all of the space between the two planes $x = 4$ and $x = -4$, as shown in Figure 46.

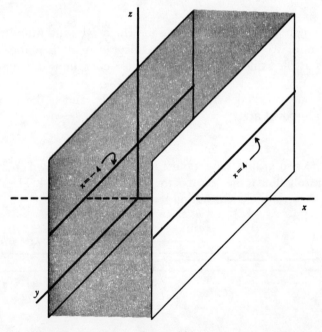

Figure 46

The intersection set of the solution sets of $y < x + 2$ and $y < -x + 2$ is graphed as the double-shaded area of Figure 47.

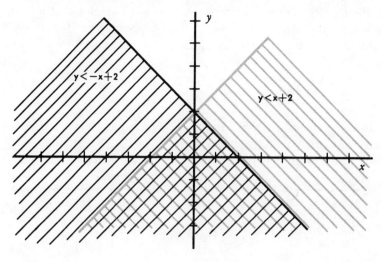

Figure 47

Sometimes two conditions may determine sets that have no elements in common. If this is the case, their intersection is called the *empty set*. One example of this is the following pair of equations in two variables:

$$y = x + 1 \text{ and } y = x + 3.$$

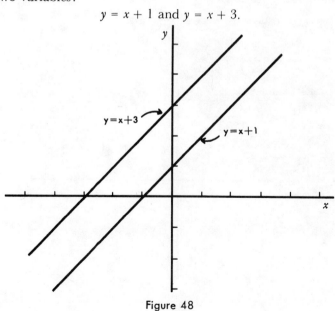

Figure 48

The graphs of the solution sets of these two equations are shown in Figure 48. Since the graphs have the same slopes, they are "going in the same direction" and will never meet. There is no number pair that will satisfy both conditions given by the two equations.

An example of an empty set that involves inequalities is

$$x > 4 \text{ and } x < -4.$$

In one dimension, the solution set of $x > 4$ is shown in Figure 49 as a solid line, and the solution set of $x < -4$ is shown by the dotted line. Note that the solution sets do not include 4 or -4.

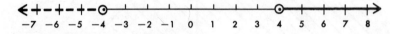

Figure 49

There is nothing in common, so the intersection is the empty set. Now consider the inequalities

$$y < x + 1 \quad \text{and} \quad y > x + 3.$$

The graphs of the solution sets of these inequalities are shown in Figure 50, and we again note that the intersection set contains no members.

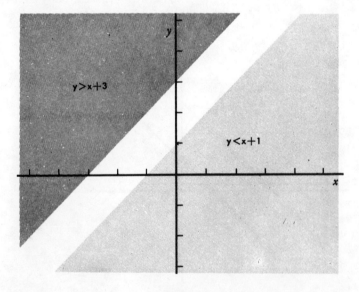

Figure 50

1. Draw graphs to find the intersection of the solution sets of the following pairs of equations and inequalities.

    a. $x > 5$ and $x < 7$

    b. $y = x + 2$ and $y = -x + 4$

    c. $y < x - 1$ and $y < 5 - 2x$

    d. $y > x + 2$ and $y < x - 2$

    e. $x^2 + y^2 < 5$ and $y > 3 - x$

2. Draw graphs of the following sets.

    a. $\{(x, y) \mid x > 2\} \cap \{(x, y) \mid y > 1\}$

    b. $\{(x, y) \mid y > x - 2\} \cap \{(x, y) \mid x > 2\}$

    c. $\{(x, y) \mid y > x^2\} \cap \{(x, y) \mid y < 1\}$

    d. $\{(x, y) \mid x \geq 0\} \cap \{(x, y) \ y \geq 0\} \cap \{(x, y) \mid x + 2y \leq 6\}$
    $\cap \{(x, y) \ 2x + y \leq 6\}$

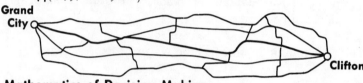

Grand City

Clifton

The Mathematics of Decision Making

Which is the most efficient way to use a certain machine? Which routes will be the best for a soft-drink company's delivery trucks? What amounts of two ingredients will produce the most profitable mixture? Questions like these, involving decisions, are often asked by business, industry, and government. In most cases, the decisions are related to money; a minimum cost or a maximum profit is desired. A new field of mathematics, called *linear programming*, has been developed to provide a mathematical basis for making decisions of the type just described. Thus, mathematics can often be used to select the best course of action from a number of possible courses. Some of the aspects of linear programming are quite complex, but it basically involves many of the ideas about equations, inequalities, graphs, and solution sets that you have worked with in this booklet. You can best understand what is involved in a linear programming problem by studying the following example.

Suppose a manufacturing plant is making two models of a product — Model I and Model II. Both models require a certain amount of time to be fabricated on two machines — Machine A and Machine B. The time in minutes required on each machine for each unit is given in Table 6.

Table 6

	Model I	Model II
Machine A	3	3
Machine B	2	5

If neither machine is scheduled to operate more than 180 minutes per day, the question then arises, "How many of each model should be manufactured per day to yield the most profit to the company if Model I produces a profit of $4 per unit and Model II produces a profit of $8 per unit?"

Let's see if we can use mathematics to reach a decision.

Let x be the number of Model I to be made each day, and y the number of Model II to be made. The profit P for a day's production can be expressed as

$$P = \$4x + \$8y. \tag{1}$$

The time limitation on Machine A can be expressed as

$$3x + 3y \leq 180, \tag{2}$$

and the time limitation on Machine B as

$$2x + 5y \leq 180. \tag{3}$$

It is also assumed that

$$x \geq 0 \text{ and} \tag{4}$$
$$y \geq 0; \tag{5}$$

that is, the number of models produced on either machine could not be a negative number.

Thus we are looking for the values of x and y that will give the largest value of P satisfying equation (1) subject to the restrictions imposed on x and y by the inequalities (2), (3), (4), and (5).

What you need to do first, then, is to obtain the intersection set satisfying conditions (2), (3), (4), and (5).

In Figures 51a and 51b, the shaded area $ABCD$ is the graph of the intersection set that contains all the possible ordered pairs that meet the conditions of the problem. We want to find some point in the graph whose coordinates, when substituted in the equation

$$P = \$4x + \$8y,$$

will give the largest possible value for P. We could just start selecting points and hope that the coordinates might be ones that would solve our problem, but this might be a tedious procedure.

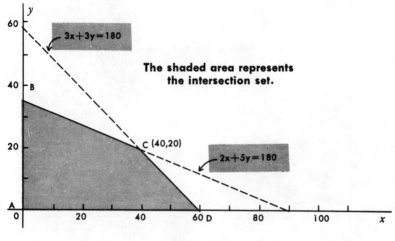

Figure 51a

Instead, let's work with the equation
$$P = \$4x + \$8y$$
and write it as
$$8y = -4x + P, \quad \text{or} \quad y = -\frac{1}{2}x + \frac{P}{8}.$$

This last equation is in the form $y = mx + b$, and so $-\frac{1}{2}$ would represent the slope of the graph of the solution set of the equation, and $\frac{P}{8}$ would represent the y intercept. The greater the y intercept of such a straight-line graph, the greater the value of P. If we draw a family of lines having the slope $-\frac{1}{2}$, as in Figure 51b, we find that the line with the greatest y intercept that also contains points from the shaded intersection set is the line k through the point C. Therefore, the coordinates of point C will give the greatest value for P when substituted in the equation $P = \$4x + \$8y$.

We see that the coordinates of C are
$$x = 40 \quad \text{and} \quad y = 20.$$
Substituting these values of x and y in (1) gives you the largest profit under the given restrictions:
$$P = \$4(40) + \$8(20)$$
$$= \$160 + \$160$$
$$= \$320.$$

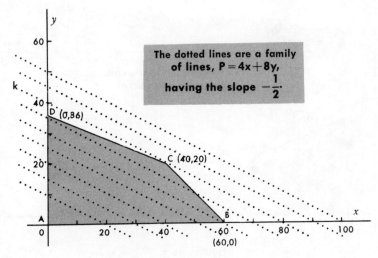

Figure 51b

For a second example, let's consider a problem in which a drug manufacturer uses either of two compounds (let's call them Alpha and Beta) or a combination of them in a special mixture to be sold as a vitamin supplement. The manufacturer wants each mixture to contain 920,000 units of vitamin A and an equal amount of vitamin D. Compound Alpha contains 184,000 units of vitamin A and 40,000 units of vitamin D per ounce, while compound Beta contains 92,000 units of vitamin A and 200,000 units of vitamin D per ounce. Alpha costs $.20 per ounce and Beta costs $.40 per ounce. The problem resolves itself into finding which combination of Alpha and Beta will give 920,000 units each of vitamins A and D, and at the same time will minimize the cost.

If we let x hold a place for the number of ounces of Alpha, and y hold a place for the number of ounces of Beta, then the following conditions must be met:

$$x \geq 0 \qquad \text{(It is obvious that there cannot}$$
$$y \geq 0 \qquad \text{be a negative number of ounces.)}$$
$$184{,}000x + 92{,}000y \geq 920{,}000 \quad \text{or} \quad 2x + y \geq 10$$
$$40{,}000x + 200{,}000y \geq 920{,}000 \quad \text{or} \quad x + 5y \geq 23.$$

The graph of the intersection set of the first four conditions is the shaded area *DEFGH*, shown in Figure 52*a*.

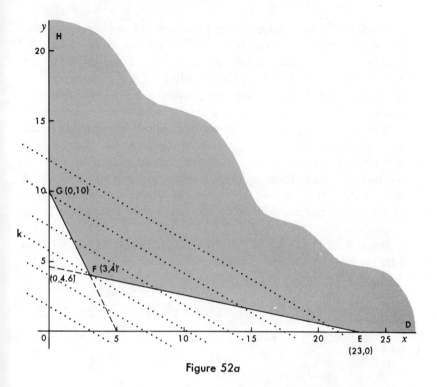

Figure 52a

The manufacturer would like to find values for x and y that would give the lowest possible cost. We can find the amounts of the compounds that would produce a minimum cost by writing the equation $C = \$.20x + \$.40y$ as $y = -\frac{1}{2}x + \frac{5}{2}C.$

We would like to find a line having a slope of $-\frac{1}{2}$ that contains a point from the intersection set graph and that cuts the y axis at the lowest possible point (thus giving a minimum value for C). By drawing a family of lines with slope $-\frac{1}{2}$ as shown in Figure 52a, we see that the line that contains a point from the intersection graph and gives the lowest possible value of C is the line k through the point $(3,4)$. The least cost will result when 3 ounces of Alpha and 4 ounces of Beta are to be used in the vitamin supplement.

We have considered two examples of problems that can be solved by linear programming methods. In both cases, we noticed that the ordered pair that met the conditions of the problem and produced either a maximum profit or minimum cost was the coordinates of a point at one of the *vertices* of the graph of the intersection set formed from the conditions of the problem. It can be shown that, in any linear programming situation, either the solution will always be found at a vertex of the intersection graph, or there will be an infinite number of solutions found along a line joining two vertices of the intersection graph. Thus, to solve a decision-type problem, it is necessary only to graph the solution sets of the mathematical conditions of the problem and then check the coordinates of points at the vertices of the intersection graph to see which ordered pair will give the desired result in a profit or cost equation.

For a final example, suppose the cost of Alpha is increased to \$.80 per ounce. The cost equation then becomes

$$C = \$.80x + \$.40y.$$

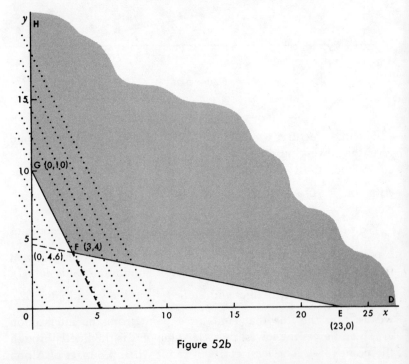

Figure 52b

From Figure 52b, we see that the coordinates of the three vertices of the intersection graph are E (23,0), F (3,4), and G

(0,10). Let's check all three of these sets of ordered pairs in the new cost equation.

$x = 23, y = 0$	$x = 3, y = 4$	$x = 0, y = 10$
$C = \$18.40 + 0$	$C = \$2.40 + \$1.60$	$C = 0 + \$4.00$
$C = \$18.40$	$C = \$4.00$	$C = \$4.00$

We see that the lowest cost occurs along the line joining vertices F and G. This is not surprising when we notice that the corresponding terms containing variables in the two equations; $C = \$.80x + \$.40y$ and $920,000 = 184,000x + 92,000y$, are in the same ratio.

We have examined two very simple problems that show how graphing methods can help make important decisions. Most of the practical problems to which linear programming is applied are much more complex than these examples. Nevertheless, you have the background for a new branch of mathematics that is becoming more important every day.

EXERCISE SET 12
Linear Programming Problems

1. Given the equation $P = x + y$ and the conditions that $x \geq 0$, $y \geq 0$, $x + 3y \leq 11$, and $3x + y \leq 9$, find values for x and y that will make P a maximum. What is the maximum value of P, subject to these restrictions?

2. Find the maximum value of $P = 2x + y$, subject to the same restrictions as those in Problem 1.

3. Given the equation $C = x + y$, subject to the conditions that $x \geq 0$, $y \geq 0$, $x + 3y \geq 11$, and $3x + y \geq 9$, find values for x and y that will make C a minimum. What is the minimum value of C, subject to these restrictions?

4. Find the minimum value of $C = 2x + y$, subject to the same restrictions as those in Problem 3.

5. A grocery store owner stocks two brands of chocolates, A and B. The store owner has room for no more than sixty boxes of chocolates in the store. He also knows that at least two times as much brand A is sold as brand B. If he makes 10¢ profit on a box of brand A and 12¢ profit on a box of brand B, how many boxes of each kind should he stock to make the maximum profit?

6. A farmer can buy two different special feed mixtures for his cattle. He wants the feed to supply at least five pounds of protein and two pounds of minerals for each cow each week. Feed K is 70% protein and 30% minerals, while feed M is 80% protein and 20% minerals. If feed K costs $1 per pound and feed M costs $.80 per pound, what ratio of feed K to feed M must he use to insure a minimum cost?

A Backward Glance and a Look to the Future

The story of graphing is only partly told in this section, for you have seen just a few beginnings, but these beginnings are basic to what can follow. The most important single idea of a graph is that it is a picture, actually a geometric picture, of a collection of points, called a *locus*, that satisfies certain conditions. For example, a circle in a plane is a set of points that is a fixed distance from a given point; a straight line has a property of constant slope. When conditions such as these are translated into the form of an equation, you can begin to see how algebra and geometry help each other. The geometry serves as a good model to show what is happening in the most abstract form of an algebraic equation.

There is just a small introduction in this section to three-dimensional graphing. We hope your appetite is whetted sufficiently in this direction to do some experimenting on your own. There are many fascinating surfaces in space like spheres, cylinders, cones, surfaces of revolution, ruled surfaces, and many more that have equations and, furthermore, can be graphed. Why not see what you can do along this line?

Then there is the lesson to be learned, that sometimes problems involve more than three variables so that three dimensions are not enough to represent the geometric interpretation. This means that man's imagination must be such that he can interpret problems involving *n* dimensions with what he already knows of the more simplified problems in one, two, or three dimensions, without actually having to draw a model.

With this, the challenge is yours to enjoy.

PART IV

Short Cuts

IN COMPUTING

Short Cuts and
Our Base 10 Numerals

The Quick Calculation Story

When Johnny was age 7, he was thrilled when his father would let him take over the lawnmower and cut a small part of the family lawn. But now Johnny is age 13, and he has acquired a thorough dislike for cutting grass.

This very short story illustrates a pattern that was followed in a much bigger story—the story of computing with numbers. When man first invented numbers and methods of counting and computing, he was fascinated with his new ideas. In many ancient lands, computing was considered such a special art that only a select few worked with numbers. As time passed, the fascination of calculation began to fade, and computing began to seem more like work than play. As man acquired more knowledge and as his life became more complex, he started to search for ways to shorten the work of computation. This led to the development of computing devices such as the abacus, logarithms, the slide rule, mechanical adding machines, and the electronic computers of today.

Although man has invented many computing devices, he has not completely eliminated the need for "good, old-fashioned" computing skills. After all, you can't always carry a calculating device with you to perform the computations necessary for checking a grocery bill, a lunch check, or a bank statement. However, there are many things that can be done to aid in mental computation and cut down on the amount of paper work needed for a calculation.

Aids to quick calculation can be classified in one of two ways: (1) as methods that enable you to use common ideas of computation in a more rapid fashion, or (2) as methods that give a completely different approach to certain types of computations. Part IV is mainly concerned with the second type of short cut listed above. Many short cuts in this second category are very special and can be used only under certain conditions. However, the more knowledge a person has about such short cuts, the more useful they become. Also, an understanding of the ideas behind a computing short cut should enable one to discover original calculation timesavers. Besides the saving of time, you should find it enjoyable to amaze your friends with the calculating powers you can gain from this part of the book.

From Counting to Rapid-Rapid Counting

Counting objects, one at a time, is an adding process—adding 1 more to the number last counted. Counting by 2's, 5's, 10's, or any other number can be used to shorten the counting process. The addition facts, such as $8 + 5 = 13$, when memorized, are also rapid ways of counting. The repeated addition of a number, such as

$$8 + 8 + 8 = 24,$$

which is memorized as $3 \times 8 = 24$, becomes one of the multiplication facts used for more rapid work.

Most quick calculating methods and computing short cuts are based upon a knowledge of the structure of our system of numeration. As you probably know, our numeration system is a base ten system, which means that we *group* by tens when counting in our system. We make use of the "ten-ness" of our numeration system when we speed up the work of adding a column of numbers by combining successive numbers that add up to 10. For example, in finding the sum of 7, 8, 2, 3, 4, 5, 1, and 7, we could think 7, 17, 20, 30, 37.

The principle of place value (which says that the value represented by a digit in a numeral depends upon the place the digit occupies in the numeral) is another important part of our numeration system. The principles of grouping by tens and place value enable us to use only ten basic symbols to represent any number. If we count ten "ones," we change it to one "ten"; ten "tens" become one "hundred"; ten "hundreds" become one "thousand"; and so on. This can be expressed in symbols as

$$(10) = 10$$
$$10(10) = 100$$
$$10(10)(10) = 1,000$$
$$10(10)(10)(10) = 10,000$$
$$\vdots$$

This pattern says that each time we multiply by 10, we increase the digit places by one. We show this by moving digits one place to the left. If a digit place is left empty, we put in a zero.

Man has even invented a special way to show this type of repeated multiplication.

You might guess ahead of time what this notation is if you notice that when 10 is a factor (factors are numbers multiplied to give a product) 4 times—

$$(10)(10)(10)(10),$$

the answer is 1 followed by 4 zeros: 10,000. The number of factors and the number of zeros are both 4. Therefore, the 4 is emphasized by placing it to the right and above the 10. Thus, $(10)(10)(10)(10)$ is written as 10^4. This 4 is called an *exponent*. The 10 is called the *base* of the exponent. You can use this method of showing repeated multiplication to write the products of various *powers* of 10.

$$10 = 10^1$$
$$(10)(10) = 10^2 = 100$$
$$(10)(10)(10) = 10^3 = 1,000$$
$$(10)(10)(10)(10) = 10^4 = 10,000$$

To complete this story, you should remember that a decimal point was invented as a symbol to mark the "ones" place. Then the place to the right of the "ones" is one tenth the size of a "one," or a "tenth"; the next place to the right is one tenth of a "tenth," or a "hundredth"; and so on. This can be expressed in symbols as

$$\frac{1}{10} = 0.1$$

$$\frac{1}{(10)(10)} = 0.01$$

$$\frac{1}{(10)(10)(10)} = 0.001$$

$$\frac{1}{(10)(10)(10)(10)} = 0.0001.$$

Exponential notation is a powerful tool that can be used to shorten the effort needed to do repeated multiplications and divisions. For example,

$$\frac{(10^5)(10^2)}{10^3}$$

may be changed to

$$\frac{[(10)(10)(10)(10)(10)][(10)(10)]}{(10)(10)(10)},$$

which reduces to

$$(10)(10)(10)(10) \text{ or } 10,000.$$

You may have noticed that the product of $(10^5)(10^2)$ has a total of 7 factors of 10 obtained by combining 5 factors from 10^5 and 2 factors from 10^2.

You can express this fact in symbols by writing

$$(10^5)(10^2) = 10^{5+2} = 10^7.$$

This example illustrates the fact that, when you multiply exponential quantities having the same base, the product will be the base with an exponent equal to the sum of the exponents of the original factors.

You may have noticed, too, that the quotient of $\frac{10^7}{10^3}$ is an expression containing only 4 factors of 10. The 3 factors of 10 in the denominator and 3 of the 7 factors of ten in the numerator divide out, leaving $(7 - 3)$ or 4 factors of 10 in the numerator. Therefore,

$$\frac{10^7}{10^3} = 10^{7-3} = 10^4.$$

In other words, when you perform a division involving exponential quantities having the same base, the quotient will be the base with an exponent obtained by subtracting the exponent of the divisor from that of the dividend.

These laws of exponents can be used to considerable advantage in shortening certain computations.

EXERCISE SET 1
Exponential Computing

Use the laws of exponents to do the following computations.

1. $(10)(10^2) =$

2. $(10^2)(10^3) =$

3. $(10^3)(10^4) =$

4. $\dfrac{10^2}{10} =$

5. $\dfrac{10^3}{10^2} =$

6. $\dfrac{10^5}{10^3} =$

7. $\dfrac{10^6}{10^6} =$

8. $\dfrac{10^4}{10^6} =$

9. $\dfrac{10^3}{10^5} =$

10. $\dfrac{(10^3)(10^4)}{10^7} =$

11. $\dfrac{(10^6)(10^6)}{(10^5)(10^4)} =$

12. $\dfrac{(10^3)(10^9)}{(10^5)(10^6)(10^4)} =$

Multiplying by 5

We have seen how a knowledge of exponential concepts and the simplicity of multiplication and division with powers of 10 can be helpful in rapid computing. Now let's see how these ideas can lead to some special short cuts in computing.

One of the short cuts that can give you more speed is one that uses the fact that 2 times 5 equals 10. How does this help? Well, it doesn't help at all unless you know, first of all, how to multiply or divide rapidly by 10, or 100, or 1,000, and so on.

As you know, the "ten-ness" of our numeration system says the following are true.

$$10 \times \quad .0032 = \quad 0.032$$
$$10 \times \quad .032 = \quad 0.32$$
$$10 \times \quad .32 = \quad 3.2$$
$$10 \times \quad 3.2 = \quad 32$$
$$10 \times \quad 32 = \quad 320$$

When you multiply by 10, you merely change the places occupied by the digits 3 and 2. This makes multiplying by 10 very simple. The short cut really amounts to moving the decimal point one place to the right for each multiplication by 10.

The next step is to show how multiplying by 5 can be simplified.

Here is how you can multiply

$$5 \times 32.$$

1. Take $\frac{1}{2}$ of 32: $\quad \frac{32}{2} = 16$

2. Multiply 16×10: $16(10) = 160$
   Therefore, $5 \times 32 = 160$.

Try it on this problem.

$$5 \times 648$$

1. $\frac{1}{2}$ of 648 $= \frac{648}{2} = 324$

2. Multiply by 10: $324(10) = 3,240$
   Therefore, $5 \times 648 = 3,240$.

If we use a symbol, such as N, as a placeholder for any number to be multiplied by 5, it is easy to show why this short cut works.

1. Start with a number: $\quad N$

2. Multiply by 5: $\quad 5N$

3. Express 5 as $\frac{10}{2}$: $\quad \frac{10}{2}N$

4. Change the order of
   multiplying and $\quad \frac{N}{2}(10)$
   dividing:

This last expression can be read as:

"Divide N by 2, and then multiply by 10."

You should notice that there isn't much advantage of this method over the ordinary multiplication process if the original number is not easily divisible by 2. This would happen if too many of the digits were odd instead of even. Thus the multiplication

$$5 \times 3,579$$

can probably be done just as quickly by the usual multiplication process as by first dividing by 2 and then multiplying by 10. Try the computation both ways to see how they compare.

Usual method:	Short-cut method:

$$
\begin{array}{r}
3579 \\
\times 5 \\
\hline
17,895
\end{array}
\qquad
\begin{array}{r}
1789.5 \\
2)\overline{3579} \\
1789.5 \times 10 = 17,895
\end{array}
$$

The use of symbols, such as N, to hold a place for any number of a certain set of numbers and to establish general mathematical relationships is a major part of the branch of mathematics called *algebra*. In this section, we shall often use the methods of algebra to show why a computational short cut works.

EXERCISE SET 2
Multiplication by 5

1. Try the following multiplications with a friend to see who can get the correct answers first. One of you multiply by 5 the regular way, and the other by the new way. You could also choose teams and see which method wins.

a. 5(46)

b. 5(68)

c. 5(842)

d. 5(1,604)

e. 5(2,632)

f. 5(8,246)

g. 5(3,206)

h. 5(28.64)

i. 5(0.0244)

2. Make a short-cut rule for multiplying by 50; by 500; by .5; by 0.05.

Multiplying by 25

It is also possible to do multiplications involving 25 by a short method. The clue to finding a short cut to multiplying by 25 comes from the fact that $25 = \dfrac{100}{4}$.

A problem like

$$25 \times 32$$

can be done mentally by doing two things:

1. Divide 32 by 4: $\dfrac{32}{4} = 8$

2. Multiply 8 by 100: $8(100) = 800$

Therefore, $25 \times 32 = 800$.

We can justify this short cut by doing a bit of very easy reasoning. The simple proof showing that this will work for any number goes like this:

$$25N = (\frac{100}{4})N = (\frac{N}{4})(100),$$

which means: Multiplying a number by 25 is the same as dividing by 4 and then multiplying by 100.

There isn't much advantage to this short cut if you can't divide by 4 easily. In a problem like

$$25 \times 3{,}579,$$

there is very little difference between the two methods.

Usual method:	Short-cut method:
3579	894.75
$\times$ 25	4)3579.00
17895	
7158	
89,475	$894.75 \times 100 = 89{,}475$

EXERCISE SET 3
Multiplication by 25

Use the short-cut method to do the following multiplications.

1. 25(16)   **2.** 25(64)

3. 25(88)   **4.** 25(48)

5. 25(124)

Dividing by 5 or 25

The short cuts for multiplying numbers by 5 or 25 suggest that there probably should be short cuts for dividing by 5 or 25. First, let's consider a problem involving a division by 5, such as

$$\frac{32}{5}.$$

It is easy to write this as

$$\frac{32(2)}{10},$$

which is

$$\frac{64}{10}$$

or

$$6.4.$$

In general, the steps taken can be expressed as

$$\frac{N}{5} = \frac{2 \times N}{2 \times 5} = \frac{2N}{10}.$$

These symbols say, "Dividing any number, N, by 5 is the same as doubling the number and then dividing by 10."

The same type of thinking is used in a short cut for dividing a number by 25. If you have the problem

$$\frac{32}{25},$$

the answer can be found by thinking

$$\frac{32}{25} = \frac{32(4)}{25(4)} = \frac{128}{100} = 1.28.$$

In general, for any number, N, this can be written as

$$\frac{N}{25} = \frac{4 \times N}{4 \times 25} = \frac{4N}{100}.$$

In words, this says, "Dividing a number, N, by 25 is the same as multiplying the number by 4 and then dividing by 100."

EXERCISE SET 4
Division by 5 and 25

Perform the indicated operations, taking advantage of short cuts.

1. $\dfrac{221}{5}$ 2. $\dfrac{221}{25}$

3. $\dfrac{308}{5}$ 4. $\dfrac{308}{25}$

5. $\dfrac{8.22}{5}$ 6. $\dfrac{8.22}{25}$

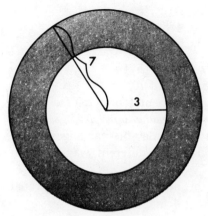

$$A = \pi 7^2 - \pi 3^2$$
$$= \pi(7^2 - 3^2)$$
$$= \pi(7 + 3)(7 - 3)$$
$$= \pi \times 10 \times 4$$
$$= 40\pi$$

$$
\begin{array}{cccc}
 & 3 & (3+4)(4+5) & 5 \\
345 \times 11 = & 3 \, , & 7 \quad 9 & 5
\end{array}
$$

A Variety
of Short Cuts

Some Important Principles of Arithmetic

We have seen how a knowledge of our numeration system and some ideas of algebra can be helpful in developing some simple computing short cuts. There are also some basic principles of arithmetic that are often helpful in developing rapid computing techniques.

If, when adding a column of figures, you encountered the combination

$$5 + 6 + 7 + 4 + 3,$$

it would be helpful to recognize this as being equivalent to

$$5 + (6 + 4) + (7 + 3),$$

giving you a simple addition of 5, 10, and 10.

This example illustrates two fundamental principles of arithmetic called the *commutative* and *associative principles of addition*, which state that the members of a sum may be rearranged (commuted) in any order and associated in different ways without changing the sum.

The commutative principle of addition can be illustrated algebraically by

$$a + b = b + a,$$

and the associative principle of addition by

$$(a + b) + c = a + (b + c).$$

There is also a *commutative principle of multiplication*. For example, you know that

$$3 \times 7 = 7 \times 3,$$

and, in general,

$$a \times b = b \times a.$$

The statement

$$(3 \times 7) \times 4 = 3 \times (7 \times 4)$$

illustrates the *associative principle of multiplication*.

Another basic principle of arithmetic, *the distributive principle*, is illustrated by the following statement:

$$7(8 + 6) = (7 \times 8) + (7 \times 6).$$

The distributive principle states that when an entire sum is multiplied by a number, the multiplier can be distributed to each member of the sum, and vice versa. This can be expressed algebraically as

$$a(b + c) = ab + ac.$$

These basic principles also play an important role in the development of numerous computing short cuts. As you study the explanations of the short cuts given in this booklet, see if you can recognize applications of these basic principles.

This section of the book contains a wide variety of short cuts, many of which are related to our numeration system, ideas of algebra, and basic principles of arithmetic.

Multiplying by 11

Multiplying by 11 is a simple process under any circumstances, but you can easily learn to find the product of 11 and any number without any paper and pencil work. Let's see how to shorten the steps by first doing a problem in the standard form. Take 254 × 11 as an example.

```
    2 5 4
×     1 1
-----------
    2 5 4      Multiplying by 1
  2 5 4        Multiplying by 10 shifts digits 1 place to the left.
-----------
  2 7 9 4
```

Notice that the first and last digits of 254 are also the first and last digits of the answer. The 9 in the answer came from adding the 4 and its left-hand neighbor 5; the 7 in the answer came from adding the 5 and its left-hand neighbor 2. So, when you have a product involving 11, just follow these steps:

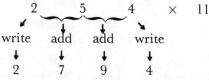

If some of the neighboring digits add up to 10 or more, you must remember to carry in the usual way, but you can do it mentally. Because of the carrying, start at the right and work to the left. For example:

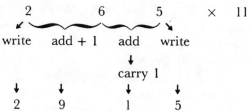

With practice, you can mentally multiply any number by 11.

EXERCISE SET 5
Finding Products Involving 11

Find the following products.

1. 352×11 **2.** 11111×11

3. 12345×11 **4.** 2178×11

5. 98765×11

 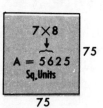

Squaring Numbers Ending in 5

Look at the following products to see if you can discover a short cut for squaring numbers ending in 5.

$$5 \times 5 = 25$$
$$15 \times 15 = 225$$
$$25 \times 25 = 625$$
$$35 \times 35 = 1225$$
$$45 \times 45 = 2025$$

$$\cdot$$
$$\cdot$$
$$\cdot$$

Can you discover a pattern that will enable you to write the answers for 55×55, 65×65, and so on?

As you study this series of products, you notice that the answers all end in 25. You might have expected this, because you are squaring numbers ending in 5.

But what about the other digits to the left of the 25 in the products? You might think, and rightly so, that the 6 in **6**25 is in some way related to the 2's in 25×25, that the 12 in **12**25 is related to the 3's in 35×35, and so on. Notice that **6** is **2** times 3 and **12** is **3** times 4. The following pattern seems to work:

When computing the product 25×25, multiply 2 (the tens digit) times the next whole number, 3. Write the answer, 6, followed by 25. Thus $25 \times 25 = 625$.

When computing the product 35×35, multiply 3 (the tens digit) times the next whole number, 4. Write the answer, 12, followed by 25. Thus, $35 \times 35 = 1225$.

When computing the product 45×45, multiply 4 (the tens digit) times the next whole number, 5. Write the answer, 20, followed by 25. Thus, $45 \times 45 = 2025$.

You can easily show why this short cut works. A two-digit number ending in 5 can be written as

$$10N + 5,$$

where N holds a place for the tens digit of the number. You want to establish that

$$(10N + 5)^2 \quad = \quad 100 \quad (N) \quad (N+1) \quad + \quad 25.$$

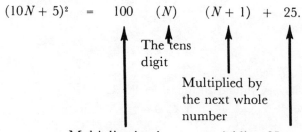

The tens digit

Multiplied by the next whole number

Multiplication by 100 puts zeros in the ones and tens places.

Adding 25 puts a 5 in the ones place and a 2 in the tens place.

To show this, you can expand the expression on the left-hand side and see if it is exactly the same as the right-hand side.

$$(10N + 5)^2 = (10N + 5)(10N + 5)$$
$$= (10N)(10N) + (5)(10N) + (5)(10N) + 25$$
$$= 100N^2 + 100N + 25$$
$$= 100N(N + 1) + 25$$

This work shows that the short cut can be used to find the square of any number having a ones digit of 5. Do you recognize the use of the distributive principle in the proof for the short cut?

EXERCISE SET 6
Squaring Practice

Use the short cut to compute the products of the following numbers ending in 5.

1. 55×55	**2.** 65×65
3. 75×75	**4.** 85×85
5. 95×95	**6.** 105×105
7. 115×115	**8.** 125×125

$$3 \times 4 \quad 3 \times 7$$
$$\downarrow \qquad \downarrow$$
$$37 \times 33 = \overset{\frown}{1 \quad 2} \quad \overset{\frown}{2 \quad 1}$$

$$7 \times 8 \quad 1 \times 9$$
$$\downarrow \qquad \downarrow$$
$$79 \times 71 = \overset{\frown}{5 \quad 6} \quad \overset{\frown}{0 \quad 9}$$

Connected by a Sum of 10

Do you notice a pattern in this series of products?

$$\begin{aligned}
2 \times 8 &= 16 \\
12 \times 18 &= 216 \\
22 \times 28 &= 616 \\
32 \times 38 &= 1,216 \\
42 \times 48 &= 2,016
\end{aligned}$$

.
.
.

This is very similar to the series

$$\begin{aligned}
5 \times 5 &= 25 \\
15 \times 15 &= 225 \\
25 \times 25 &= 625 \\
35 \times 35 &= 1,225 \\
45 \times 45 &= 2,025
\end{aligned}$$

.
.
.

In the first series, with a 2 and an 8 for the ones places, the product ends in 16. In the second series, with a 5 in each ones place, the product ends in 25. In both series, the other digits seem to follow the same pattern.

Thus, in the first series, 16 apparently comes from 2×8, and in the second series, 25 comes, as we have learned earlier, from 5×5. But why do the other digits follow the same pattern?

Perhaps there is some relationship between 2 and 8 that is also true of 5 and 5. An obvious connection is that the sums are the same:

$$2 + 8 = 10 \text{ and } 5 + 5 = 10.$$

Let's try to give a more general explanation of this situation.

If $(10N + M)$ is any two-digit number, with N the tens digit and M the ones digit, then $[10N + (10 - M)]$ will be a corresponding two-digit number with the same tens digit and in which the sum of its ones digit and the ones digit of the other number will always be 10.

We need to show that the product of the two numbers
$$(10N + M)[10N + (10 - M)]$$
is the same as

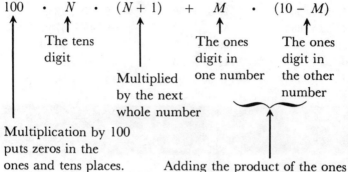

100   •   N   •   (N + 1)   +   M   •   (10 − M)

The tens digit

Multiplied by the next whole number

The ones digit in one number

The ones digit in the other number

Multiplication by 100 puts zeros in the ones and tens places.

Adding the product of the ones digits puts that product in the ones and tens places of the result.

We can use algebraic methods to change the form of $(10N + M)$ $[10N + (10 - M)]$ as follows:

$$(10N + M)[10N + (10 - M)]$$
$$= (10N + M)(10N) + (10N + M)(10 - M)$$
$$= 100N^2 + 10MN + 100N - 10MN + 10M - M^2$$
$$= 100N^2 + 100N + 10M - M^2$$
$$= 100N(N + 1) + M(10 - M)$$

This shows that $(10N + M)[10N + (10 - M)]$ is equal to $100 \cdot N \cdot (N + 1) + M \cdot (10 - M)$ for all values of M and N.

This short cut can be applied to a wide variety of products, such as

$$23 \times 27 = 6 \ 21$$
$$2 \times 3 \quad 3 \times 7$$

and

$$64 \times 66 = 42 \ 24$$
$$6 \times 7 \quad 4 \times 6$$
$$71 \times 79 = 56 \ 09$$
$$7 \times 8 \quad 1 \times 9$$

It is required that the tens digit be the same in the two numbers being multiplied and that the ones digits have a sum of 10.

Use the short cut to find the following products.

1. 52×58	**2.** 62×68	**3.** 72×78
4. 82×88	**5.** 92×98	**6.** 43×47
7. 53×57	**8.** 63×67	**9.** 73×77
10. 51×59	**11.** 37×33	**12.** 81×89
13. 54×56	**14.** 96×94	**15.** 84×86

$$43^2 = 100(43 - 25) + (50 - 43)^2 = 1800 + 49 = 1849$$
$$87^2 = 200(87 - 50) + (100 - 87)^2 = 7400 + 169 = 7569???$$

Squaring a Number Between 25 and 50

It is often advantageous to be able to find the square of a number mentally. This short cut involves squaring a number between 25 and 50, such as 46.

To find the value of 46×46, take these steps:

1. Find the difference between 46 and 25: $46 - 25 = 21$

2. Multiply this difference, 21, by 100: $21 \times 100 = 2,100$

3. Find the difference between 50 and 46: $50 - 46 = 4$

4. Square the 4 obtained in step 3: $4 \times 4 = 16$

5. Add 2,100 obtained in step 2 and 16 obtained in step 4: $2,100 + 16 = 2,116$

This is the answer:

$$46 \times 46 = 2,116$$

Let's explain this short cut by translating these steps into the language of algebra.

Use N to represent any number between 25 and 50.

1. Find the difference between N and 25: $N - 25$

2. Multiply this difference by 100: $100(N - 25)$

3. Find the difference between 50 and N: $50 - N$

4. Square the result obtained in step 3: $(50 - N)^2$

5. Add the results obtained in steps 2 and 4: $100(N - 25) + (50 - N)^2$

The result in step 5 should be the same as N^2. To show this, we simplify $100(N - 25) + (50 - N)^2$. This becomes

$100N - 2,500 + 2,500 - 100N + N^2$, or N^2.

EXERCISE SET 8
More Squaring Practice

Use the preceding short cut to find the following products.

1. 41×41	**2.** 42×42
3. 43×43	**4.** 44×44
5. 45×45	**6.** 47×47
7. 48×48	**8.** 49×49

Multiplying Two Teen Numbers

A short cut to multiplying two numbers between 10 and 20 can be illustrated by the following example:

To multiply 17×19,

1. Add 17 and 9: $17 + 9 = 26$
2. Multiply this sum by 10: $10(26) = 260$
3. Multiply 7×9: $7(9) = 63$
4. Add 260 from step 2
and 63 from step 3: $260 + 63 = 323$

$$17 \times 19 = 323$$

We can use algebra to show why this trick works.
Let one teen number be $10 + a$
and the other be $10 + b$.
Then the product can be written as

$(10 + a)(10 + b) = 10^2 + 10a + 10b + ab = 10[(10 + a) + b] + ab$.

In the above example, $a = 7$ and $b = 9$. Then, $10 + a = 10 + 7 = 17$. Now let's follow the development of the steps of the short cut to see why they lead to the correct answer.

1. $(10 + a) + b = 17 + 9 = 26$
2. $10[(10 + a) + b] = 10(26) = 260$
3. $ab = 7(9) = 63$
4. $10[(10 + a) + b] + ab = 260 + 63 = 323$

EXERCISE SET 9
More Product Practice

Use the method on page 19 to multiply any two teen numbers. With practice you should be able to do these without paper and pencil.

1. 16×13 **2.** 18×15

3. 14×17 **4.** 19×19

Multiplying Two Numbers That Differ by Two

Let's see if you can discover a pattern in the following multiplications.

$(1)(3) = 3$		$(2)(2) = 4$
$(2)(4) = 8$		$(3)(3) = 9$
.		.
.		.
.		.
$(9)(11) = 99$		$(10)(10) = 100$
.		.
.		.
.		.
$(19)(21) = ?$		$(20)(20) = 400$
.		.
.		.
.		.
$(29)(31) = ?$		$(30)(30) = 900$
.		.
.		.
.		.

All of the products in the first column are made up of factors that differ by 2. In the second column, the factors being multiplied together are equal and lie between the two factors given in the first column. The answers in the right-hand column are 1 more than the answers in the left-hand column. In most cases, it is easier to square a number than it is to find the product of two different numbers. (And you have short cuts for finding the squares of certain numbers.) The above pattern seems to indicate that to find the product of two numbers that differ by 2, you can find the square of the number between them and subtract 1.

Looking back at the columns, you can now fill in the question mark for

$$(19)(21) = ?$$

because the answer is $(20)(20) - 1 = 400 - 1 = 399.$
The answer for $(29)(31)$ is $(30)(30) - 1 = 900 - 1 = 899.$

If two numbers differ by 2, one can be represented algebraically as $(N - 1)$ and the other as $(N + 1)$. Then the statement

$$(N - 1)(N + 1) = N^2 - 1$$

is a proof for the short cut, for N represents the whole number between $N - 1$ and $N + 1$.

EXERCISE SET 10
Still More Product Practice

Use the above short-cut method to compute the following products.

1. $(39)(41)$	**2.** $(49)(51)$
3. $(59)(61)$	**4.** $(69)(71)$
5. $(79)(81)$	**6.** $(89)(91)$
7. $(14)(16)$	**8.** $(24)(26)$

Multiplying Two Numbers Just Over 100

Here is an easy short cut for multiplying two numbers between 100 and 110.

Suppose the two numbers to be multiplied are 106 and 108.

Put the numbers down in one column:	Subtract 100 from each:
106	$106 - 100 = 6$
108	$108 - 100 = 8$

Add diagonally: $106 + 8 = 114$
(The two numbers of the other diagonal would add up the same: $108 + 6 = 114.$)

Multiply the 6 and the 8: $(6)(8) = 48$
The product is

$$(106)(108) = 100(114) + 48 = 11,448.$$

Algebra can be used to prove this short cut. Start with two numbers, N and M, in the first column.

N Subtract 100 from N: $N - 100$

M Subtract 100 from M: $M - 100$

Add diagonal terms:

$$(N + M - 100)$$

Multiply $(N - 100)(M - 100)$

The answer should be

$$100(N + M - 100) + (N - 100)(M - 100).$$

Multiplying by 100 puts zeros in the ones and tens places so that the final answer will contain the digits of this product.

This final expression should prove to be NM, the product of the two original numbers, and it does after simplification.

The steps of simplifying go like this:

$100(N + M - 100) + (N - 100)(M - 100)$

$= 100N + 100M - 10{,}000 + NM - 100N - 100M + 10{,}000$

$= NM.$

EXERCISE SET 11
Product of Numbers Over 100

Use the short cut to multiply the following numbers.

1. $(104)(107)$ 2. $(108)(103)$
3. $(109)(109)$ 4. $(105)(105)$
5. $(104)(106)$ 6. $(101)(109)$

$$85 \times 91 = 100(85 - 9) + 15 \times 9 = 7600 + 135 = 7735$$

$$73 \times 96 = 100(73 - 4) + 27 \times 4 = 6900 + 108 = 7008$$

Multiplying Two Numbers Just Under 100

This short cut can save you time if you are multiplying two numbers between 90 and 100.

Suppose you wish to find the product of 93 and 96.

Put the numbers down in one column:	Subtract each from 100:
93	$100 - 93 = 7$
96	$100 - 96 = 4$

Subtract diagonals: $93 - 4 = 89$
(The other two numbers would give the same result: $96 - 7 = 89$)
Multiply the 7 and 4: $7(4) = 28$
The product is $(93)(96) = 100(89) + 28 = 8,928$.

We can easily show why this short cut works. Start with two numbers, N and M, in the first column.

N Subtract N from 100: $100 - N$
M Subtract M from 100: $100 - M$
Subtract diagonals: $N - (100 - M)$
Multiply $(100 - N)(100 - M)$

The answer should be

$$100[N - (100 - M)] + (100 - N)(100 - M).$$

Multiplying by 100 puts zeros in the ones and tens places so that they will contain the digits of this product.

This final expression should prove to be NM, the product of the two original numbers, and it does if you simplify it.

$100[N - (100 - M)] + (100 - N)(100 - M)$
  $= 100(N - 100 + M) + 10,000 - 100N - 100M + NM$
  $= 100N - 10,000 + 100M + 10,000 - 100N - 100M + NM$
  $= NM$

EXERCISE SET 12
Products of Numbers Just Under 100

Use the short cut to multiply the following numbers.

1. $(94)(97)$	**2.** $(98)(93)$
3. $(99)(99)$	**4.** $(95)(95)$
5. $(94)(96)$	**6.** $(91)(99)$

Short Cuts
with Nines

Adding Nines Is Easy

As previous sections have shown, the "ten-ness" of our numeration system has a great deal to do with the development of many computing short cuts. The close relationship between 9 and 10, illustrated by the fact that 9 is 1 less than 10, or 99 is 1 less than 100, or 999 is 1 less than 1,000, and so on, leads to several new computational short cuts.

For example, a quick way to add 9 to any number is to add 10 and subtract 1. Thus,

$$9 + 8 = 10 + (8 - 1) = 10 + 7 = 17.$$

Notice how the commutative and associative principles are used.

A good way to generalize this rule is to use N to represent any number. Then, for any number, N,

$$9 + N = 10 + (N - 1).$$

You can add 99 to any number by adding 100 and subtracting 1. Thus,

$$99 + 68 = 100 + (68 - 1) = 100 + 67 = 167,$$

and, in general, for any number, N,

$$99 + N = 100 + (N - 1).$$

Exactly the same steps are followed for adding 999—add 1,000 and then subtract 1. For example,

$$999 + 375 = 1,000 + (375 - 1) = 1,000 + 374 = 1,374,$$
and for any number, N,
$$999 + N = 1,000 + (N - 1).$$

A Trick with Nines

The short-cut method for adding a number made up of a sequence of 9's serves as the basis for a good number trick.

Ask someone to write any four-digit number: 6,921

Ask someone else to write another four-digit number: 4,317

You write: 5,682

Ask someone to write another four-digit number: 8,692

You write: 1,307

Before you ask anyone to total this column of figures, you write the answer,

$$26,919,$$

on the back of a piece of paper. Then have everyone in the audience find the total. To their surprise, your answer is correct. They will undoubtedly wonder how you were able to write the answer so quickly.

Let's examine the problem by assigning a letter to each number.

$$a = 6,921$$
$$b = 4,317 \quad a, b, \text{ and } d$$
c and e were $c = 5,682 \quad$ were numbers chosen

numbers chosen $d = 8,692 \quad$ by someone else.

by you. $e = \underline{1,307}$

$$26,919$$

There is a special way you must choose c and e. Do you see how b and c or d and e are related?

The digits in c are chosen by subtracting each digit in b from 9. The digits in e are chosen by subtracting each digit in d from 9. This process is called finding the complements of 9.

This special way of choosing c and e will guarantee that

$$b + c = 9,999$$

and

$$d + e = 9,999.$$

The sum of all five numbers looks like this:

$$a + b + c + d + e = a + (b + c) + (d + e)$$
$$= a + (9,999) + (9,999)$$
$$= a + (10,000 - 1) + (10,000 - 1)$$
$$= 20,000 + (a - 2).$$

This tells you that the sum can be found by subtracting 2 from the first number and then adding that result to 20,000. Thus, in our example, it was necessary to subtract 2 from 6,921 to get 6,919, and then place a 2 in front of 6,919 to represent the addition of 20,000 and get the correct answer, 26,919.

EXERCISE SET 13
Working with Nines

1. Develop short-cut rules for adding 8, or 98, or 998, and so on.

2. Develop short-cut rules for adding 7, or 97, or 997, and so on.

3. As a variation of the parlor trick with the four-digit numbers, find the complement of a instead of b. The problem would look like this.

6,921	a	chosen by audience
4,317	b	chosen by audience
3,078	c	complement of a
8,692	d	chosen by audience
1,307	e	complement of d
24,315	=	$20,000 + b - 2$

Multiplying Nines

Since multiplication is repeated addition, you might expect short cuts in multiplications involving 9, or 99, or 999, and so on, as well as in adding. Look at the following statement.

$$9 \times 8 = (10 \times 8) - 8$$

This says, "Nine times a number is the same as 10 times the number if you then subtract the number."

In general, for any number, N,

$$9N = (10 - 1)N = 10N - N.$$

Multiplying by 99 is quite similar, because $99 = (100 - 1)$. Therefore,

$$99 \times 46 = (100 - 1)46 = 4,600 - 46 = 4,554.$$

In general, for any number, N,
$$99N = (100 - 1)N = 100N - N.$$
Multiplying by 999 is just as easy.
$$999 \times 8 = (1,000 - 1)8 = 8,000 - 8 = 7,992,$$
and, for any number, N,
$$999N = (1,000 - 1)N = 1,000N - N.$$
You can use this principle to verify that the following number pattern is correct.
$$1 \times 9 + 2 = 11$$
$$12 \times 9 + 3 = 111$$
$$123 \times 9 + 4 = 1111$$
$$1234 \times 9 + 5 = 11111$$
The method of verification is general enough to apply to any of the steps. Try it on
$$1234 \times 9 + 5 = 11111.$$
Change 9 to $(10 - 1)$. Then the expression becomes
$$1234(10 - 1) + 5 = (1234)10 - 1234 + 5$$
$$= 12340 - 1234 + 5 = 12345 - 1234 = 11111.$$
See if you can recognize when the distributive, commutative, and associative principles are used.

Since this analysis could have been applied to any of the steps, we can conclude that this pattern will hold until you reach
$$12345678 \times 9 + 9.$$
What happens after that?

EXERCISE SET 14
Product Problems with Nines

1. Develop short-cut rules for multiplying by 8, or 98, or 998, and so on.

2. Verify that the following pattern is correct.
$$1 \times 90 + 21 = 111$$
$$12 \times 90 + 31 = 1111$$
$$123 \times 90 + 41 = 11111$$
$$1234 \times 90 + 51 = 111111$$
$$\vdots$$

Subtracting Nines

When you subtract a 9, you can more easily subtract 10 first and then add 1. For example,

$$26 - 9 = 26 - 10 + 1 = 16 + 1 = 17.$$

In general, for any number, N, we have

$$N - 9 = N - (10 - 1) = (N - 10) + 1.$$

When you subtract 99, you can think, "Subtract 100, then add 1."

$$256 - 99 = (256 - 100) + 1 = 156 + 1 = 157$$

In general, for any number, N,

$$N - 99 = (N - 100) + 1.$$

The pattern is now well established. Subtracting 999 from 3,526 can be done as

$$3,526 - 999 = (3,526 - 1,000) + 1 = 2,526 + 1 = 2,527,$$

and, in general, for any number, N, you can write

$$N - 999 = (N - 1000) + 1.$$

This principle is used quite often in making change. If you hand the clerk a $5 bill for a purchase totaling $1.95, you can think,

$$\$5 - \$1.95 = \$5 - \$2 + \$.05 = \$3.05, \text{ the correct change.}$$

EXERCISE SET 15
Difference Problems with Nines

1. Develop short-cut rules for subtracting 8; 98; 998; and so on.

2. Develop short-cut rules for subtracting 7; 97; 997; and so on.

3. Use mental arithmetic to complete the following change table.

Amount of Cash	Amount of Sale	Change
$ 5	$ .95	
$10	$ 5.95	
$20	$16.95	
$20	$ 6.95	

So far you have seen the 9's play some very special roles in adding, multiplying, and subtracting. But what about division? Well, there are some very interesting results with division, too. You can see the effects very quickly by performing the following steps.

Find the decimal equivalent of $\dfrac{1}{9}$.

Find the decimal equivalent of $\dfrac{1}{99}$.

Find the decimal equivalent of $\dfrac{1}{999}$.

You should obtain the following results.

$$\frac{1}{9} = 0.111111\overline{1} \ldots$$

$$\frac{1}{99} = 0.0101010\overline{01} \ldots$$

$$\frac{1}{999} = 0.001001\overline{001} \ldots$$

The line above the digits indicates the digits that are repeated over and over again.

Notice what happens when you take a one-digit number and multiply it by $\dfrac{1}{9}$.

$$1 \times \frac{1}{9} = \frac{1}{9} = 1 \times 0.111 \ldots = 0.11\overline{1} \ldots$$

$$2 \times \frac{1}{9} = \frac{2}{9} = 2 \times 0.111 \ldots = 0.22\overline{2} \ldots$$

$$3 \times \frac{1}{9} = \frac{3}{9} = 3 \times 0.111 \ldots = 0.33\overline{3} \ldots$$

$$4 \times \frac{1}{9} = \frac{4}{9} = 4 \times 0.111 \ldots = 0.44\overline{4} \ldots$$

$$5 \times \frac{1}{9} = \frac{5}{9} = 5 \times 0.111 \ldots = 0.55\overline{5} \ldots$$

The repeating decimal repeats the one-digit number. This pattern suggests that, if you have a repeating decimal *repeating a single digit*, you can immediately write a fraction that corresponds to the repeating decimal. Thus, as an illustration, if you have the repeating decimal

$$0.77\overline{7} \ldots,$$

where 7 is the repeated digit, then this can be written as the fraction

$$\frac{7}{9}.$$

Next consider

$$\frac{1}{99} = 0.0101\overline{01} \ldots$$

and multiply by one- or two-digit numbers.

$$2 \times \frac{1}{99} = \frac{2}{99} = 2 \times .010101 \ldots = 0.0202\overline{02} \ldots$$

$$3 \times \frac{1}{99} = \frac{3}{99} = 3 \times .010101 \ldots = 0.0303\overline{03} \ldots$$

$$4 \times \frac{1}{99} = \frac{4}{99} = 4 \times .010101 \ldots = 0.0404\overline{04} \ldots$$

$$\cdot$$
$$\cdot$$
$$\cdot$$

$$10 \times \frac{1}{99} = \frac{10}{99} = 10 \times .010101 \ldots = 0.1010\overline{10} \ldots$$

$$\cdot$$
$$\cdot$$
$$\cdot$$

$$25 \times \frac{1}{99} = \frac{25}{99} = 25 \times .010101 \ldots = 0.2525\overline{25} \ldots$$

$$\cdot$$
$$\cdot$$
$$\cdot$$

$$43 \times \frac{1}{99} = \frac{43}{99} = 43 \times .010101 \ldots = 0.4343\overline{43} \ldots$$

$$\cdot$$
$$\cdot$$
$$\cdot$$

Thus, any two-digit cycle of a repeating decimal can be written immediately as a fraction having 99 as the denominator and the two-digit number in the cycle as the numerator. The repeating decimal $.6262\overline{62}$ . . . is $\frac{62}{99}$.

You can probably guess by now what happens to the decimal representation of fractions with 999 as a denominator. There is a three-digit cycle.

$$\frac{1}{999} = 0.001001\overline{001} \ldots$$

$$\frac{2}{999} = 0.002002\overline{002} \ldots$$

$$\frac{3}{999} = 0.003003\overline{003} \ldots$$

.
.
.

$$\frac{10}{999} = 0.010010\overline{010} \ldots$$

.
.
.

$$\frac{100}{999} = 0.100100\overline{100} \ldots$$

.
.
.

$$\frac{232}{999} = 0.232232\overline{232} \ldots$$

.
.
.

$$\frac{547}{999} = 0.547547\overline{547}$$

This suggests immediately how you can write a fraction that is the equivalent of a three-digit cycle repeating decimal. The repeating decimal $0.876876\overline{876}$ . . . is $\frac{876}{999}$.

The process of changing repeating decimals to common fractions is often useful in computing, for it is usually easier to perform a multiplication or division with a common fraction than with a repeating decimal.

It is easy to do any division in which the divisor is made up of 9's. For example, to do the division $\dfrac{5,643}{999}$, it is only necessary to recognize that $\dfrac{1}{999} = 0.001001\overline{001} \ldots$ and then multiply $0.001001\overline{001}$ ... by 5,643. This multiplication follows a definite pattern and is very easy to do.

$$
\begin{array}{r}
5,643 \\
\times\ 0.001001001 \ldots \\
\hline
5643 \\
5643 \\
5643 \\
\hline
5.648648648 \ldots \\
\uparrow
\end{array}
$$

There would be an 8
in this digit position, for a
5 would be generated by
the next multiplication
of 001. The "648 pattern"
would go on indefinitely.

EXERCISE SET 16
Division Problems with Nines

1. Find the repeating decimal equivalents of the following.

a. $\dfrac{3}{9}$ or $\dfrac{1}{3}$ b. $\dfrac{24}{99}$ or $\dfrac{8}{33}$ c. $\dfrac{37}{999}$ or $\dfrac{1}{27}$

d. $\dfrac{370}{999}$ or $\dfrac{10}{27}$ e. $\dfrac{141}{999}$ or $\dfrac{47}{333}$ f. $\dfrac{39}{9,999}$ or $\dfrac{13}{3,333}$

2. Find the fractions, reduced to lowest terms, which have the following decimal equivalents.

a. $0.6666\overline{6} \ldots$ b. $0.60606\overline{0} \ldots$ c. $0.600600\overline{600} \ldots$

d. $0.27272\overline{7} \ldots$ e. $0.270270\overline{270} \ldots$

3. Use a short-cut method on the following divisions.

a. $\dfrac{352}{99}$ b. $\dfrac{4,266}{999}$

Short Cuts
for Testing and Checking

The Rules of Divisibility

Here is a problem involving the multiplication of fractions:

$$\frac{336}{13} \times \frac{143}{7}.$$

You know, of course, that this computation can be performed by finding the product of 336 and 143 and dividing that result by the product of 7 and 13. However, this would be too long a method of computation, for the work can be simplified by "dividing out" equal factors in the numerators and denominators of the fractions. For example, 336 is divisible by 7, giving 48, and 143 is divisible by 13, giving 11. (When we say 336 is divisible by 7, we mean that there is some *whole number* which when multiplied by 7, will give 336.) Thus, the problem can be reduced to the easy product 11 × 48.

Of course, it was necessary to recognize that 336 is divisible by 7 and that 143 is divisible by 13 in order to shorten the computation. Since it is not always easy to recognize whether or not one number is divisible by another, mathematicians have developed some short cuts to test for divisibility. The use of these short cuts can save you a lot of time.

Here are a few divisibility tests.

Divisibility by 2. You no doubt are already familiar with the idea that you can tell at a glance that *a number is divisible by 2 if the number ends in any one of the digits 0, 2, 4, 6, or 8.*

Divisibility by 4. *A number is divisible by 4 if the last two digits form a number that is divisible by 4.* Thus, 1,702,582 is *not* divisible by 4 because 82 (the number formed by the last two digits) is not divisible by 4. The number 1,702,584 is divisible by 4 because 84 is divisible by 4.

Divisibility by 8. *A number is divisible by 8 if the number formed by the last three digits is divisible by 8.* To find out whether or not 1,702,584 is divisible by 8, you need not divide 1,702,584 by 8, but, instead, only divide 584 by 8. By dividing 584 by 8, you find that it gives an even answer of 73. Therefore, the original number 1,702,584 is divisible by 8.

The proof of these three tests for divisibility can be handled together, because they are closely related.

Consider a four-digit number expressed in the form

$$1,000a_3 + 100a_2 + 10a_1 + a_0,$$

where a_0 represents the units digits, a_1 the tens digit, and so on. By the distributive principle, this can be changed to the following form to check for divisibility by 2:

$$1,000a_3 + 100a_2 + 10a_1 + a_0 = (100a_3 + 10a_2 + a_1)10 + a_0.$$

If the right-hand side is divisible by 2, the left-hand side will also be divisible by 2. But, since the expression $(100a_3 + 10a_2 + a_1)10$

on the right-hand side is divisible by 2 (it has a factor of 10 which is divisible by 2), the only way for the right-hand side to be divisible by 2 is for the remaining portion, a_0, to be divisible by 2. Therefore, if the ones digit is 0, 2, 4, 6, or 8, the original number is divisible by 2.

To prove the rule for divisibility by 4, use the distributive and associative principles to make the following change in form:

$$1{,}000a_3 + 100a_2 + 10a_1 + a_0 = (10a_3 + a_2)100 + (10a_1 + a_0).$$

The argument is the same as before. If the right-hand side is divisible by 4, the left-hand side will also be divisible by 4. Since the expression $(10a_3 + a_2)100$ is divisible by 4, because it has a factor of 100 which is divisible by 4, then the only way for the right-hand side to be divisible by 4 is for the remaining portion $(10a_1 + a_0)$ to be divisible by 4. The expression $(10a_1 + a_0)$ represents the number formed by the last two digits of the original number. Therefore, we see that the rule is correct.

To prove the rule for divisibility by 8, make the following change in form:

$$1{,}000a_3 + 100a_2 + 10a_1 + a_0 = a_3(1{,}000) + (100a_2 + 10a_1 + a_0).$$

The argument again repeats itself. The left-hand side is divisible by 8 if the right-hand side is. Since $a_3(1{,}000)$ is divisible by 8, because 1,000 is divisible by 8, then the expression $(100a_2 + 10a_1 + a_0)$ being divisible by 8 would guarantee that the original number meets the same condition.

The expression $(100a_2 + 10a_1 + a_0)$ represents the number formed by the last three digits of the original number, and so the rule has been established. The proof of these three divisibility rules would follow the same pattern regardless of the number of digits you might have in the original number.

Divisibility by 5. This rule is easy and is well-known. *A number is divisible by 5 if it ends in either 5 or 0.* Develop a proof for this rule similar to the one given for divisibility by 2.

Divisibility by 10. *A number is divisible by 10 if the number ends in 0.* See if you can develop a proof similar to the one given for divisibility by 2.

There are other tests of divisibility that are not as well known, but they are valuable because they are not too difficult to use and can save time in certain situations. The next few paragraphs give some of these divisibility tests.

Divisibility by 9. Let's start a multiplication table of 9's:

$$1 \times 9 = 9$$
$$2 \times 9 = 18$$
$$3 \times 9 = 27$$
$$4 \times 9 = 36$$
$$5 \times 9 = 45$$

$$\cdot$$
$$\cdot$$
$$\cdot$$

Notice that, in every case, the digits in the product add up to 9. Is this true of any multiplication fact of 9? Let's try another one:

$$11 \times 9 = 99.$$

Here the sum of the digits is a multiple of 9: $2 \times 9 = 18$. It is interesting to notice further that the digits in 18 add up to 9: $1 + 8 = 9$.

This observation leads to a general rule for divisibility by 9:

Any number is divisible by 9 if the sum of its digits is 9 or a multiple of 9.

Thus, 217,683 is divisible by 9 because the sum of the digits,

$$2 + 1 + 7 + 6 + 8 + 3,$$

is 27, which is a multiple of 9. Of course, you can check on the fact that 27 is a multiple of 9 by showing that $2 + 7 = 9$.

You don't need to add these digits and get the answer of 27 if, as you mentally add the digits, you can drop each sum of 9. Thus, when adding the digits of 217,683, you could think: $2 + 1 + 7 = 10$, drop 9, carry 1; $1 + 6 + 8 = 15$, drop 9, carry 6; $6 + 3 = 9$.

The rule for divisibility by 9 isn't difficult to prove. Any three-digit number can be written in the form:

$$100a_2 + 10a_1 + a_0,$$

and the following changes in the form can then be made:

$$(100a_2 + 10a_1 + a_0) = 99a_2 + a_2 + 9a_1 + a_1 + a_0$$
$$= (99a_2 + 9a_1) + (a_2 + a_1 + a_0).$$

If the right side of this equality statement is divisible by 9, then the left side will be divisible by 9. Since $(99a_2 + 9a_1)$ is divisible by 9, we can definitely say that if $(a_2 + a_1 + a_0)$, representing the sum of the digits of the original number, is divisible by 9, then $(100a_2 + 10a_1 + a_0)$, the original number, is also divisible by 9. We can also reverse this reasoning and say that if a number is divisible by 9, then the sum of the digits of the number is divisible by 9.

If $(a_2 + a_1 + a_0)$ is less than 9, it is the remainder obtained when the original number is divided by 9. If $(a_2 + a_1 + a_0)$ is greater than 9, but not a multiple of 9, the above process could be repeated and would finally produce a number less than 9 that would represent the remainder obtained when the original number is divided by 9.

Divisibility by 3. *Any number is divisible by 3 if the sum of the digits is divisible by 3.*

Thus, 12,345 is divisible by 3 because $1 + 2 + 3 + 4 + 5$ is divisible by 3. Since 9 is divisible by 3, you can cast out 9's just as you did in the test for divisibility by 9. However, if the final remainder is a 3, 6, or 9, the original number will be divisible by 3.

See if you can prove this rule in the same way that you proved the rule for divisibility by 9.

Divisibility by 7. *Any number is divisible by 7 if the difference between twice its units digit and the number formed by the other digits is exactly divisible by 7.*

Is 336 divisible by 7? The test shows that it is. You double the units digit 6, and get 12, and then subtract 12 from 33 and get 21. Since 21 is divisible by 7, the original number, 336, is divisible by 7. This test is more of a novelty than a practical short cut. Nevertheless, it is interesting to know that there is such a test, and it can come in handy in particular situations.

The proof would go something like this. Take a three-digit number as an illustration:

$$100a_2 + 10a_1 + a_0,$$

and make the following changes:

$$(100a_2 + 10a_1 + a_0) = 100a_2 + 10a_1 + 20a_0 - 20a_0 + a_0$$
$$= 21a_0 + 10(10a_2 + a_1 - 2a_0).$$

An important step was the addition and subtraction of $20a_0$. Now, if the right-hand side of the above equality expression is divisible by 7, then the left-hand side will be divisible by 7. Since $21a_0$ is divisible by 7, the right-hand side of the expression will be divisible by 7 if $(10a_2 + a_1 - 2a_0)$ is divisible by 7. Thus, we have shown that, if the difference between twice the units digit and the number formed by the other digits of a number is divisible by 7, the original number is divisible by 7.

Divisibility by 11. *Any number is divisible by 11 if the sum of the digits in the odd places minus the sum of the digits in the even places is 0 or a number divisible by 11.*

Thus, 6,523 is divisible by 11 because

$6 + 2 = 8$ (The sum of the digits in the

even places: **6** 5 **2** 3)   — 2nd digit

— 4th digit

$5 + 3 = 8$ (The sum of the digits in the

odd places: 6 **5** 2 **3**)   — 1st digit

— 3rd digit

and the difference, $8 - 8$, is divisible by 11.

The proof of this is along the same lines as the previous ones. You can use a three-digit number as an example.

Change the original number,

$$100a_2 + 10a_1 + a_0,$$

to the form

$$100a_2 + 10a_1 + a_0 + 100a_1 - 100a_1 + 100a_0 - 100a_0$$
$$= (110a_1 - 99a_0) + 100(a_2 - a_1 + a_0).$$

If $(a_2 - a_1 + a_0)$ is divisible by 11, we must accept the conclusion that the original number, $(100a_2 + 10a_1 + a_0)$, is divisible by 11, and vice versa, because $(110a_1 - 99a_0)$ is divisible by 11.

Divisibility by 13. *Any number is divisible by 13 if the sum of 4 times the units digit and the number formed by the other digits is divisible by 13.*

Is 143 divisible by 13? The test for divisibility says:

Multiply 4 times 3: $4 \times 3 = 12$

Add this product to 14: $12 + 14 = 26.$

Since 26 is divisible by 13, 143 is also divisible by 13.

This test, like the one for divisibility by 7 is mostly a novelty rather than a short cut. Nevertheless, it can be practical in some situations.

See if you can prove this law of divisibility by using the following equality statement:

$$100a_2 + 10a_1 + a_0 = 100a_2 + 10a_1 + a_0 + 40a_0 - 40a_0$$
$$= 10[(10a_2 + a_1) + 4a_0] - 39a_0.$$

EXERCISE SET 17
Working with Divisibility

1. Use the tests for divisibility to find the factors of the following numbers.

a. 111   b. 1,530   c. 855

d. 4,554   e. 2,088   f. 1,612

2. What values of the digit d will make the numbers divisible by 2?

a. 821d b. 82d1

3. What values of the digit d will make the numbers divisible by 3?

a. 821d b. 82d1

4. What values of the digit d will make the numbers divisible by 5?

a. 821d b. 82d1

5. What values of the digit d will make the numbers divisible by 9?

a. 821d b. 82d1

6. What values of the digit d will make the numbers divisible by 11?

a. 821d b. 82d1

Checking by Casting Out Nines

When the test for divisibility by 9 was explored, it was pointed out that, when a number is divided by 9, the remainder is closely related to the sum of the digits of the number.

For example, when you divide 36 by 9 you get a remainder of 0, and when you add the digits in 36 and cast out 9 you also have 0. If you divide 56 by 9 you get a remainder of 2, and when you add the digits in 56, you get $5 + 6 = 11$. Casting out 9 will give you $11 - 9 = 2$. Divide 152 by 9. You will find that there is a remainder of 8. Now add the digits: $1 + 5 + 2$. The sum is also 8. Any time the sum of the digits exceeds 9, cast out 9's or multiples of 9 and keep what is left. This will be the remainder when the original number is divided by 9. This method of "casting out 9's" applies to any number no matter how many digits it has.

The method of "casting out 9's" can be used as an easy and rapid check for addition, subtraction, and multiplication. It can be used to check division, too, but the process becomes more complicated because of the many steps in a division problem.

Let's look at some examples that show how this principle can be used to check computations. Here is an example that shows how addition can be checked by casting out 9's.

```
  3 6 5    Add the digits and cast out 9's:    3 + 6 + 5 ──► 5
  2 8 3                                         2 + 8 + 3 ──► 4
+ 1 5 7                                         1 + 5 + 7 ──► 4
─────────
  8 0 5
```

Add the digits in this sum and cast out 9's: $8 + 0 + 5$ ──► 4

Add these and cast out 9's: $5 + 4 + 4$ ──► 4

The fact that these answers are the same indicates the sum is probably correct.

Let's try to show why this check works. Remember that, when you add the digits and cast out 9's, you are finding the remainder obtained from a division by 9. Since you are concerned with division by 9, you can write 365 as $(40 \times 9) + 5$; 283 as $(31 \times 9) + 4$; and 157 as $(17 \times 9) + 4$. The 5, 4, and 4 are the remainders when the original numbers are divided by 9. The sum of the original numbers can now be written as

$$(40 \times 9) + 5 + (31 \times 9) + 4 + (17 \times 9) + 4,$$

and, using the commutative law of addition, you can write
$365 + 283 + 157 = (40 \times 9) + (31 \times 9) + (17 \times 9) + 5 + 4 + 4$.
The sum of the remainders, $5 + 4 + 4$, can be written as $(1 \times 9) + 4$, the 4 representing the remainder when this sum of remainders is divided by 9. Now you can write the original addition problem,

$$365 + 283 + 157,$$

as

$$(40 \times 9) + (31 \times 9) + (17 \times 9) + (1 \times 9) + 4,$$

and, since $[(40 \times 9) + (31 \times 9) + (17 \times 9) + (1 \times 9)]$ is divisible by 9, the 4 also represents the remainder when the original number is divided by 9.

This same type of reasoning could have been used with any other numbers in the original sum. Thus, you see that the remainder when the sum of several numbers is divided by 9 is the same as the remainder when the sum of the remainders (obtained by dividing the original number by 9) is divided by 9.

Of course, there could be a possible error in the original sum that would not be detected by the casting-out-9's check if you made an error that caused the answer to differ from the correct answer by 9 or a multiple of 9.

The same method also works with subtraction. Here is a typical example:

$$8\ 7\ 5\ 6$$

Add the digits and cast out 9's: $8 + 7 + 5 + 6 \longrightarrow 8$

$$-\ 1\ 3\ 8\ 7$$

$1 + 3 + 8 + 7 \longrightarrow 1$

$$\overline{7,3\ 6\ 9}$$

Subtract these and cast out 9's:

Add the digits in this answer and cast out 9's:

$8 - 1 \longrightarrow 7$

$7 + 3 + 6 + 9 \longrightarrow 7$

The fact that these answers are the same indicates the subtraction is likely to be correct.

The explanation of why the check works is closely related to the one given for addition. See if you can supply the explanation.

The check of a multiplication problem goes like this:

$$2\ 8\ 7$$

Cast out 9's in the sum of the digits: $2 + 8 + 7 \rightarrow 8$

$$\times\ 1\ 6\ 5$$

$1 + 6 + 5 \rightarrow 3$

$$\overline{1\ 4\ 3\ 5}$$

$$1\ 7\ 2\ 2$$

Multiply these and cast out 9's:

$$\underline{2\ 8\ 7}$$

$$\overline{4\ 7,3\ 5\ 5}$$

$3 \times 8 = 24$

$2 + 4 = 6$

Cast out 9's in the answer:

$4 + 7 + 3 + 5 + 5 \longrightarrow 6$

The fact that these answers agree is strong evidence that the product is probably correct.

Let's try to explain the casting-out-9's check for multiplication. Instead of giving this explanation for the two numbers in the example, let a and b hold a place for any two factors of a product. When a is divided by 9, let's represent the quotient with m and the remainder with x, and when b is divided by 9, let's represent the quotient with n and the remainder with y. Then

$$a = 9m + x$$

and

$$b = 9n + y.$$

The product ab can now be written in the following way:

$$ab = (9m + x)(9n + y)$$
$$= (9m)(9n) + (9m)y + (9n)x + xy.$$

Since $(9m)(9n) + (9m)y + (9n)x$ is divisible by 9, the remainder when ab is divided by 9 is the same as the remainder when xy is divided by 9.

EXERCISE SET 18
Using Casting Out 9's

Perform the indicated operations and check the answers by the method of casting out 9's.

1. $\begin{array}{r} 1\,8,2\,5\,6 \\ +\quad 2,9\,5\,4 \\ \hline \end{array}$

2. $\begin{array}{r} 7,2\,6\,5 \\ -\quad 4\,8\,2 \\ \hline \end{array}$

3. $\begin{array}{r} 8,7\,6\cdot 6 \\ \times\quad 1\,8\,2 \\ \hline \end{array}$

A Backward Glance and a Look to the Future

You can see that there are many applications of the variety of short cuts presented in this book. Some of the short cuts presented are quite special and only work when certain conditions are satisfied, such as the squaring of a number ending in 5. But others are very general and apply to all numbers, such as the method of casting out 9's to check computation. In any event, short cuts are very intriguing—some because of their novelty, others because of their practicality.

Of course, this book contains only a small sample of methods and ideas that can be used to speed up computation. There are many other ways to shorten calculation time. This book presents ideas that point out why the short cuts work and how they were developed. Some of these ideas can be used to discover new computing short cuts. You might be able to discover some important new ideas that would help speed up calculations. Why not try?

Extending Your Knowledge

If you would like to know more about short cuts in computing, the following books will give you more ideas:

MEYERS, LESTER, *High Speed Math*. D. Van Nostrand Company, 1947

STICKER, HENRY, *How to Calculate Quickly*. Dover Publications, 1955

PART V

Computing
Devices

Some Old,
But Useful, Computers

The Computing Story

Which of the following calculations would you rather do?

$$8 + 5 + 7$$
$$16,825 \times 96,703$$
$$\frac{5,297 \times 8,261}{6,394}$$

You probably selected the first problem, for you can do this calculation in your head. You could do the second and third problems with pencil and paper, but you would probably be quite happy if you could find some sort of device to do the calculating for you. Your wish for a computing aid would not be new; through the centuries, man has searched for devices, machines, and ideas that would reduce the time and energy needed for burdensome calculations, and thus provide more time to develop ideas that could build a richer life.

The history of computing devices is a story of man's desire to do difficult calculations in a small amount of time. Man has been so successful in fulfilling these desires that all kinds of problems which once seemed impossible now can be solved because machines are able to do many computations in a fraction of the time it would take most human beings to do them. In fact, some problems had never even been tried until recent years because the task was too tedious.

Part V tells the story of many computing devices. In a sense it is a partial history of mathematics itself, for it includes one of the most ancient devices, the abacus, and one of the most modern, the electronic digital computer. The contrast between these two is almost unbelievable.

Of course, man would have had no use for any kind of computing device if he had not invented a system of numeration to be used in counting and calculating. Some of man's first efforts at counting were probably marks scratched on a smooth surface, or notches cut in a stick, with a mark or notch matched with each object being counted. As civilizations grew more complex, and it became necessary to count larger numbers, the marks or notches became cumbersome. Man then developed the idea of using different symbols (numerals) and combinations of symbols to represent different counts (numbers). For example, if □ represented ten objects and ∠ represented four objects, then □ ∠ could be used to represent fourteen objects.

Some peoples included ideas of *grouping* and *place value* in their numeration systems. These ideas can be illustrated by our present numeration system. To represent the number of marks in Figure 1, we would group them in tens as shown in the figure and then write the number 23, representing two groups of ten marks plus three ones. The placement of the symbols 2 and 3 determines their value.

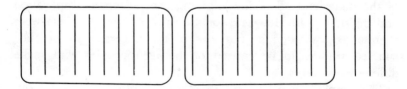

Figure 1

Our numeration system is called the Hindu-Arabic system. It is thought that it originated in India in the third century B. C. and was carried to Baghdad in the eighth century A. D. From there it found its way to Europe. Through the centuries there have been changes in the symbols used in the system, but the main ideas used in the system have not changed. There was no zero in the system at first. The origin of the zero concept is uncertain, but the earliest inscription that has been found in India containing a symbol as a placeholder for no counters was dated 876 A. D. The principles of grouping, place value, and zero formed the basis for a numeration system in which computations could be easily performed.

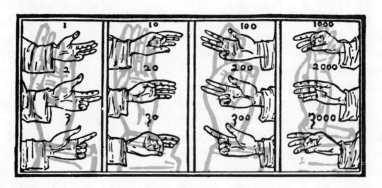

Finger Computing

Even before numeration systems were fully developed, man made use of computing devices. The fingers served as one of the earliest computing aids.

You probably don't think of your fingers as a computing device, but you undoubtedly used them as counters during your early childhood. Of course, this didn't last long after you knew the order of the digits and all of the addition, subtraction, multiplication and division facts. However, for several centuries, numbers were represented by different positions of the fingers. This practice enabled trading to be carried on between peoples who spoke different languages.

The simplest type of computing was counting on the fingers. Methods were also found to help the illiterate in their multiplication of numbers larger than 5. For example, to multiply 6 × 8, they raised 1 finger on the left hand (the difference between 6 and 5), and raised 3 fingers on the right hand (the difference between 8 and 5). The total number of fingers raised was $(1 + 3) = 4$. This was to be the number of "tens" in the answer. They then counted the number of fingers not raised on each hand; 4 on the left and 2 on the right. The product of these two numbers, $4 \times 2 = 8$, was the number of "ones" in the answer. Therefore, $6 \times 8 = 48$.

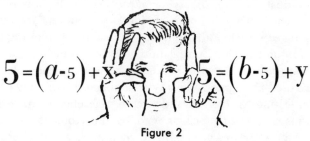

$$5 = (a-5) + x \qquad 5 = (b-5) + y$$

Figure 2

We can show why this method works by using some ideas of algebra. Suppose we want to find the product of any two numbers between 5 and 10. Let's use the letters a and b to hold places for the two numbers. Then $(a - 5)$ and $(b - 5)$ will represent the number of fingers up on each hand when the finger method of computing is used. We know that the number of fingers up plus the number down on each hand is equal to 5, so if we let x hold a place for the number of fingers down on one hand and y a place for the number down on the other, we get the relationships

$$5 = (a - 5) + x \text{ and } 5 = (b - 5) + y,$$
$$\text{or } 10 - a = x \text{ and } 10 - b = y.$$

Since $(a - 5)$ and $(b - 5)$ represent the number of fingers *up* on each hand and $(10 - a)$ and $(10 - b)$ represent the number *down* on each hand, the finger method of computing says that we can find the product ab by evaluating

$$10[(a - 5) + (b - 5)] + (10 - a)(10 - b).$$

This reduces to

$$10 (a + b - 10) + (100 - 10a - 10b + ab),$$
or
$$10a + 10b - 100 + 100 - 10a - 10b + ab,$$

and finally to just

$$ab.$$

Thus we see why the finger computing method works.

The Counting Boards

Before the development of writing materials such as papyrus, parchment, and paper, sharp-pointed sticks were used to mark numbers on tables covered with sand or dust. These dust-covered tables were used to record counts and aid in simple computations. Historians have traced the word "abacus" to the Greek word for dust. The dust counting board or *abacus* eventually was replaced by a table with lines ruled on it having loose counters arranged on the lines to keep track of the counting. This type of counting device was in common use in Europe until the beginning of the seventeenth century A. D. In other parts of the world, beads placed on rods were used as counters. This type of abacus is still in use today.

In the writings of Horace and Cicero, reference is made to the three forms of counting boards used by the Romans: a grooved table with beads, a marked table for counters, and the primitive dust board. The counters were originally made of stone, and this explains the origin of the word "calculate," which comes from the word "*calculus*" meaning pebble. From this you can also see the interesting connection between the words calcium, talc, and chalk, for the pebbles or *calculi* were cut out of limestone, a calcium compound used in talc and chalk.

The Chinese used bamboo rods as counters as early as 542 B. C., and developed the modern form of the abacus by the twelfth century A. D. Even today, one type of abacus is in universal use in China in banks and shops of all kinds. Those who use them in the business world are as efficient with them as an expert typist is with a typewriter. Some people are so efficient with the abacus that they can compute more rapidly than people who operate adding machines. Figure 3 shows a Chinese abacus. Each upper bead has a value of 5 of its order and each lower bead a value of 1 of its order. The number represented on the abacus is 3,075.

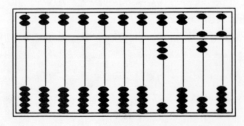

Figure 3

The Japanese are known to have used an abacus as early as the sixteenth century, and it is still in universal use in Japan at the present time. Their abacus is called a *soroban*. A computation contest was held recently between an American using an adding machine and a Japanese using the soroban, and the Japanese won.

There are two types of sorobans found in Japan today, a five-bead and a four-bead. The four-bead (four beads below the crossbar) is illustrated in Figure 4 because it has become quite popular in recent years.

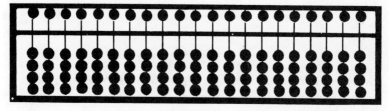

Figure 4

The technique in using this soroban is to move beads with the thumb and forefinger. The bead above the crossbar has a value of 5 of its order. Each bead below the crossbar has a value of 1 of its order. The vertical rods indicate different place values. If you are dealing only with whole numbers, the first rod on the right can stand for "ones," the next rod to the left, "tens," and so on. Beads brought toward the crossbar give the digits for each place value. For example, the settings in Figure 5 represent the digits shown at the tops of the various rods.

1 2 3 4 5 6 7 8 9

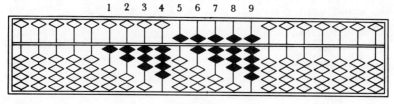

Figure 5

To add with a soroban, you move the appropriate beads toward the crossbar; to subtract, you move them away. For example, suppose you want to add 652 and 87. You could set 652 on the abacus as shown in Figure 6a. Then add 87 by first moving a five and two one counters to the crossbar on the "ones" rod. To add eight tens, move one of the hundreds beads (10 tens) to the bar, move the 5 tens bead away $(10 - 5 = 5)$, and move three of the tens beads to the crossbar $(5 + 3 = 8)$. The result, 739, is shown on the soroban in Figure 6*b*.

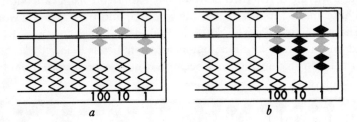

Figure 6

Multiplication can be done on the soroban as repeated additions. Suppose you want to find the product of 3,914 and 274. This product could be written as

$$3,914 \times 274 = 3,914(200 + 70 + 4).$$

If we apply the multiplier 3,914 to each part of the indicated sum (the distributive principle), we have

$$3,914 \times 274 = (3,914 \times 200) + (3,914 \times 70) + (3,914 \times 4).$$

By rearranging these products we get

$$3,914 \times 274 = (2 \times 391,400) + (7 \times 39,140) + (4 \times 3,914).$$

Therefore, to find the product of 3,914 and 274, 3914 is put on the soroban four times, 39140 seven times, and 391400 twice, the addition of these thirteen values being cumulative. In other words, we would use the soroban to add 3,914 four times, 39,140 seven times, and 391,400 twice.

Division can be done on the soroban by making repeated subtractions of the divisor from the dividend, the number of subtractions being counted on an unused rod. The number of subtractions made would be the quotient.

Of course, if you know the basic multiplication and division facts, you can do parts of the computations in your head and eliminate some of the additions or subtractions. With practice, an abacus can be a great time saver when you are doing calculations.

You can work with decimal fractions on an abacus by locating the "ones" position at some other vertical rod. If the "ones" position is located on the third rod from the right on the abacus pictured in Figure 7, the number represented is 372.08.

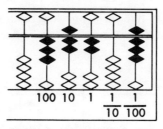

Figure 7

The most popular abacus of Western Europe during the Middle Ages was a table with lines ruled horizontally. Counters were placed on the lines and in the spaces between the lines.

In England they called an abacus of this type a "counting table." Only one counter was placed in a space, or four on a line. If five counters were on a line, then four were removed and one "carried" to the space above. If two counters appeared in a space, one was removed and one was "carried" to the line above. This explains the use of the word "carry" in addition vocabulary.

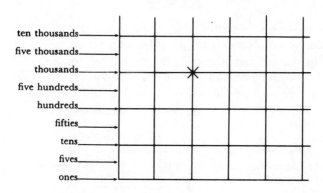

Figure 8

The "thousands" lines were marked with a cross so that the eye could spot the correct lines more easily. We do the same type of thing when we use commas to separate numerals into groups of three. Figure 8 shows the typical marking of a counting board.

Arithmetic books in the 15th and 16th centuries showed drawings of the line abacus to illustrate how to add and subtract.

Figure 9 shows an example of adding three numbers on a counting board. Each number is displayed by counters in a separate compartment. The final sum is shown in the compartment to the right. Note that the counters in the final answer came from applying the rule that five on a line are replaced by one on the space above and that two in a space are replaced by one on the line above.

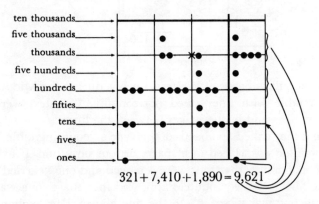

$$321 + 7,410 + 1,890 = 9,621$$

The last column gives the total of the other three columns.

Figure 9

Figure 10 is an example of subtracting two numbers on a counting board. Notice that, in subtracting, you take away from the first column what you see in the second column and show what is left in the third column. Since there is not a "fifty" counter in the first column, it is necessary to replace a "hundred" counter with two fifties in order to do the subtraction.

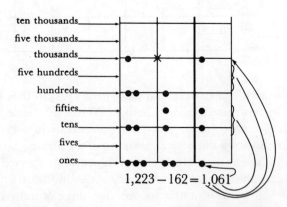

$$1,223 - 162 = 1,061$$

The third column expresses the difference between the counters in the first and second columns.

Figure 10

EXERCISE SET 1
Working with the Abacus

1. Construct a rod abacus using a $1'' \times 1'' \times 12''$ piece of wood, a $1'' \times 12''$ strip of heavy cardboard, seven long nails, and thirty-five $1'' \times 1''$ cardboard squares. Locate seven equally spaced points $1\frac{1}{2}''$ apart along the middle of one side of the board and along the middle of the cardboard strip. Punch holes in the center of each cardboard square. Put one square on each nail, force each nail through the cardboard strip at a marked point, place four more squares on each nail, and pound the nails into the board at marked points. A completed abacus is shown below.

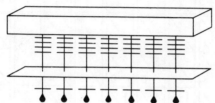

2. Use your abacus to do the following computations.
a. $284 + 756$	b. $90.6 + 964.3$	c. $7.84 - 5.32$
d. $925 - 338$	e. 6×23	f. $374 \times 5,638$

3. Construct a line abacus or "counting table" similar to the one shown in Figure 8. Use the abacus to do the following computations.
a. $243 + 312$	b. $609 + 18 + 224$	c. $416 + 8,305 + 98$
d. $368 - 256$	e. $7,318 - 693$	f. $4,723 - 3,947$

Multiplying with Bones

In 1617, John Napier, a Scottish mathematician, published a book called *Rabdologia*. The name "Rabdologia" was used by the author to refer to a method that simplified the tedious work of multiplication. This method made use of a set of numerating rods, sometimes called "bones." John Napier did not reap a great profit from this invention, for he died in 1617, the year in which the *Rabdologia* was published. However, his methods and devices for aiding in computations are still well known today.

Figure 11 shows a set of rods or bones similar to the ones Napier used for multiplication. There is a rod for each of the ten digits, and each rod contains the basic multiplication facts for that digit. There is also an index rod listing the digits 1 through 9.

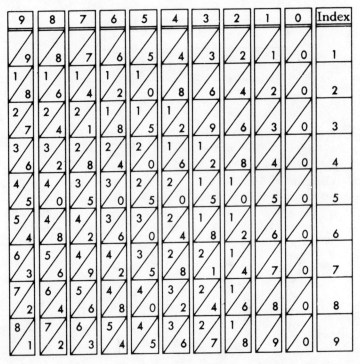

Figure 11

The rods can be a great help in multiplication. For example, to multiply 3,654 by 8, select the rods for 3, 6, 5, and 4 and place them together in the order of the digits in the number, 3,654 as shown in Figure 12. Then, by locating the eighth row of multiplication facts with the help of the index rod, the product of 3,654 and 8 can be obtained quickly. Merely bring down the first digit of the eighth row and then add the remaining figures diagonally to get the rest of the digits in the product. If the sum of two "diagonal" digits is greater than 9, bring down the first digit of the sum and carry 1 to the next digit place.

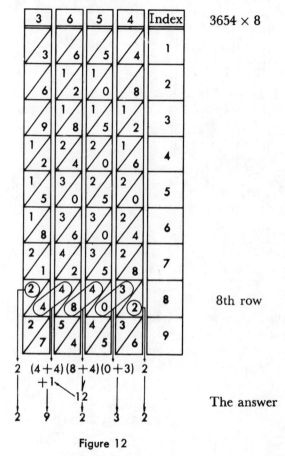

3654×8

8th row

The answer

Figure 12

This is much easier and faster than ordinary multiplication, because the facts are already written down. All you need to do is add one number in the upper triangle on one rod to the number in the lower triangle of the next rod to its left. Thus multiplication in this form requires only simple addition — a real advantage.

The index makes it much easier to solve a problem like (3,654) (237). Select the rods for 3, 6, 5, and 4 of the first number and place them together in the correct order, with the index to the right. Then find the products of 2(3,654), 3(3,654), and 7(3,654) as shown in Figure 13 on the next page.

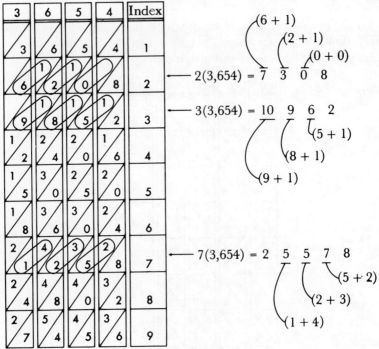

Figure 13

But the product of 3,654 and 237 is $(200 + 30 + 7)(3,654)$ or $(200 \times 3,654) + (30 \times 3,654) + (7 \times 3,654)$, so it is necessary to change the products obtained with the bones to the proper place values and then add the products.

$$
\begin{aligned}
7(3,654) &= 25,578 \\
30(3,654) &= 109,620 \\
200(3,654) &= 730,800 \\
\hline
237(3,654) &= 865,998
\end{aligned}
$$

EXERCISE SET 2
Using Napier's Rods

1. Make a set of Napier's rods on cardboard. You will need duplicate rods for each digit repeated. Remember, too, to make rods for the zero multiplication facts, because they have *place* value.

2. Use your set of Napier's rods to perform the following multiplications.

    a. $(365)(24)$ b. $(275)(381)$ c. $(2,075)(381)$

    d. $(2,075)(3,081)$ e. $(323)(99)$ f. $(293)(392)$

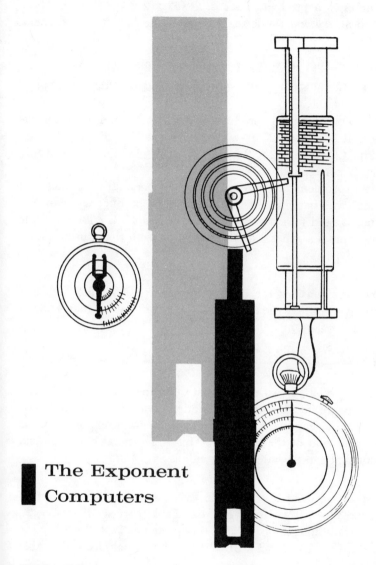

The Exponent Computers

A Table of Twos

John Napier worked with many computing devices and methods other than his computing rods. Let's try to rediscover one of his most important ideas by working with the familiar number 2.

If we raise 2 to various powers, we will get the results shown in Table 1. In the table, the symbol 2^4 tells you to use 2 as a factor four times in a product. That is, $2^4 = 2 \times 2 \times 2 \times 2 = 16$. The 4 is called an *exponent*, while the 2 is called the *base* of the exponent.

Table 1

2^1	2^2	2^3	2^4	2^5	2^6	2^7	2^8	2^9	2^{10}	2^{11}	2^{12}	2^{13}	2^{14}
2	4	8	16	32	64	128	256	512	1024	2048	4096	8192	16384

Let's work with the product of 8×16. By multiplication we know that $8 \times 16 = 128$. We also know that $8 = 2^3$, $16 = 2^4$, and $2^3 \times 2^4 = (2 \times 2 \times 2)(2 \times 2 \times 2 \times 2) = 2^{3+4} = 2^7$. As we would expect, 2^7 corresponds to 128 in Table 1.

In this problem, we multiplied two exponential quantities that had the same base, and we found that the product was equal to that base raised to a power equal to the *sum of the exponents* of the factors. This is a general mathematical concept that can be stated with symbols as

$$a^m \times a^n = a^{m+n}.$$

Now let's consider the division problem $4,096 \div 512$. By long division we can find the quotient to be 8. But we see that $4,096 = 2^{12}$, $512 = 2^9$, and $2^{12} \div 2^9 =$

$$\frac{2 \times 2 \times 2 \times 2 \times 2 \times 2 \times 2 \times 2 \times 2 \times 2 \times 2 \times 2}{2 \times 2 \times 2 \times 2 \times 2 \times 2 \times 2 \times 2 \times 2} = 2^{12-9} = 2^3 = 8.$$

Here we divided two exponential quantities that had the same base, and we found that the quotient was equal to the base raised to a power equal to the *difference* between the exponent of the divisor and that of the dividend. This is a general mathematical concept that can be stated with symbols as

$$a^m \div a^n = a^{m-n}$$

Suppose we want to square 32. By long multiplication we find that $32 \times 32 = 1,024$. However, from Table 1 we see that $32 = 2^5$, and know that squaring 2^5 is interpreted as $2^5 \times 2^5 = (2 \times 2 \times 2 \times 2 \times 2)(2 \times 2 \times 2 \times 2 \times 2) = 2^{2 \times 5} = 2^{10}$, which, from the table is 1,024.

When an exponential quantity is raised to a power, the result is equal to the base raised to a power equal to the *product* of the exponents involved. This can be stated in a general form as

$$(a^m)^n = a^{mn}.$$

Finally, let's do a square root problem. Suppose we want to find the square root of 1,024, that is, a number multiplied by itself to give 1,024. From Table 1 and a knowledge of the meaning of exponents we know that $1,024 = 2^{10}$, and that

$$\sqrt{2^{10}} = \sqrt{2 \times 2 \times 2 \times 2 \times 2 \times 2 \times 2 \times 2 \times 2 \times 2} = 2^{10 \div 2} = 2^5 = 32.$$

This illustrates the important mathematical idea that, when a root of an exponential power is to be computed, the result is equal to the base raised to a power equal to the original exponent *divided* by the index of the root. (The index of a square root is 2, of a cube root, 3, and so on.) This can be stated in symbols as

$$\sqrt[n]{a^m} = a^{m \div n}.$$

The ideas presented in the above examples illustrate methods that can be used to do some special computations in a rapid fashion. First of all, let's construct a new and simpler table from Table 1. In Table 2, the base 2 has been eliminated and only the exponents and the results when 2 is raised to each exponent are presented.

Table 2

Exponent of 2	1	2	3	4	5	6	7	8	9	10	11	12	13	14
Number	2	4	8	16	32	64	128	256	512	1024	2048	4096	8192	16384

We can use this table for certain computations. For example, if we wish to multiply 256 by 32, we find the exponent for 32 (5), the exponent for 256 (8), *add* the two exponents (5 + 8 = 13), and then find the number that corresponds to the exponent 13 (8,192). Thus $32 \times 256 = 8,192$.

If we want to divide 16,384 by 256, we find the exponent for 16,384 (14), *subtract* from it the exponent for 256 (14 − 8 = 6), find the number corresponding to an exponent of 6, and obtain 64 as the result. Thus, $16,384 \div 256 = 64$.

To find the square root of 4,096, we find its exponent (12), *divide* it by 2 (12 ÷ 2 = 6), and get the answer 64 from Table 2. Thus $\sqrt{4096} = \sqrt{2^{12}} = 2^6 = 64$.

To compute 8^3 with the table, we merely find the exponent for 8 (3), *multiply* it by 3 (3 × 3 = 9), and read the number that corresponds to the exponent 9 (512) as the result.

Thus, $8^3 = (2^3)^3 = 2^9 = 512$.

Table 2 certainly can be a time saver for some particular computations. With it, certain multiplications can be performed by merely finding simple sums; divisions can be reduced to simple subtraction problems; raising a number to a power can be handled by a simple multiplication; and square roots can be found by dividing by 2.

Use Table 2 to find answers to the following problems.

1. 8(128)   **2.** (128)(32)   **3.** $\dfrac{2,048}{128}$

4. $\dfrac{512}{32}$ **5.** 64^2 **6.** 16^3 **7.** 4^5

8. $\sqrt{4,096}$ **9.** $\sqrt{256}$ **10.** $\sqrt[3]{4,096}$

$$2.15 \times 10^3$$

$$1.8567 \times 10^3$$

$$3.6 \times 10^3 \times 2.4 \times 10^1 \times 3.65 \times 10^2 \times 1.86 \times 10^5$$

The Power of Powers of Ten

The methods of the preceding section are quite useful if we are computing with numbers that are powers of 2. However, Table 2 would be of no value in computations such as 19.3×7.6, 5^7, or $\sqrt{13.4}$. For exponents to be a great aid in computation, it is necessary that one be able to write all numbers as a power of some one base. We are faced with the problem of finding a number that could be used as a base in a complete computing system based on exponents. This was the same problem encountered by John Napier when he started his search for computing aids in the latter part of the sixteenth century. Since the Hindu-Arabic numeration system uses grouping by tens, Napier eventually worked with 10 as a base for a system of exponent computing. Let's examine Napier's discoveries by considering some patterns involving the number 10.

Every time a number is multiplied by 10, the digits move one place to the left. If we start with 2 and multiply by 10, we must move the 2 from the "ones" place to the "tens" place. If we continue to multiply by 10, each step looks like this:

$$
\begin{aligned}
&\text{Start with:} && 2 \\
&\text{Multiply 2 by 10:} && 2(10) &&= &&20 \\
&\text{Multiply 20 by 10:} && 20(10) &&= &&200 \\
&\text{Multiply 200 by 10:} && 200(10) &&= &&2000
\end{aligned}
$$

$\vdots$

Multiplying a number by 10 doesn't change the digits — it just changes their position.

The following pattern —

$$
\begin{aligned}
10 &= && 10 \\
(10)(10) &= && 100 \\
(10)(10)(10) &= && 1000 \\
(10)(10)(10)(10) &= && 10000
\end{aligned}
$$

$\vdots$

— leads to the exponent form

$$
\begin{aligned}
10^1 &= && 10 \\
10^2 &= && 100 \\
10^3 &= && 1000 \\
10^4 &= && 10000
\end{aligned}
$$

$\vdots$

The exponent is a way of showing how many factors of 10 are being multiplied.

Now let's examine some division relationships involving powers of 10. One way to evaluate

$$
\frac{10^4}{10^6}
$$

is to change it to the form

$$
\frac{10 \times 10 \times 10 \times 10}{10 \times 10 \times 10 \times 10 \times 10 \times 10}
$$

or

$$
\left(\frac{10}{10}\right)\left(\frac{10}{10}\right)\left(\frac{10}{10}\right)\left(\frac{10}{10}\right)\frac{1}{10 \times 10},
$$

which becomes

$$
\frac{1}{10^2}
$$

But if we attempt to evaluate $\dfrac{10^4}{10^6}$ by using the rule of exponents,

$$\frac{a^m}{a^n} = a^{m-n},$$

we have

$$\frac{10^4}{10^6} = 10^{4-6} = 10^{-2}.$$

In order for negative exponents to make sense and be consistent with the rules of exponents, we must define 10^{-2} to be the same as $\dfrac{1}{10^2}$ or 10^2 to be the same as $\dfrac{1}{10^{-2}}$, or, in general,

$$10^{-m} = \frac{1}{10^m} \text{ or } 10^m = \frac{1}{10^{-m}}.$$

This also leads to the interesting product,

$$10^m \times 10^{-m} = 10^m \times \frac{1}{10^m} = 1.$$

If we use the rule of exponents,

$$a^m \times a^n = a^{m+n},$$

then we get

$$10^m \times 10^{-m} = 10^{m-m} = 10^0.$$

Since $10^m \times 10^{-m}$ is known to be 1, it necessarily follows that 10^0 should be defined as 1.

This development gives a rather complete picture of numbers expressed with powers of 10. Table 3 shows an interesting pattern.

Table 3

10^{-3}	10^{-2}	10^{-1}	10^0	10^1	10^2	10^3
$\dfrac{1}{10^3}$	$\dfrac{1}{10^2}$	$\dfrac{1}{10^1}$	$\dfrac{1}{10^0}$	$\dfrac{1}{10^{-1}}$	$\dfrac{1}{10^{-2}}$	$\dfrac{1}{10^{-3}}$
$\dfrac{1}{10 \times 10 \times 10}$	$\dfrac{1}{10 \times 10}$	$\dfrac{1}{10}$	1	10	10×10	$10 \times 10 \times 10$
$\dfrac{1}{1000}$	$\dfrac{1}{100}$	$\dfrac{1}{10}$	1	10	100	1000
.001	.01	.1	1.	10.	100.	1000.

Since our numeration system is based on the principle of grouping by tens, any number can be expressed as the product of a number between 1 and 10 and a power of 10. This is illustrated by the following examples.

$$.048 = \frac{4.8}{100} = 4.8 \times 10^{-2}$$

$$.48 \ = \frac{4.8}{10} = 4.8 \times 10^{-1}$$

$$4.8 \ = 4.8 \times 1 = 4.8 \times 10^{0}$$

$$48 \ = 4.8 \times 10 = 4.8 \times 10^{1}$$

$$480 \ = 4.8 \times 100 = 4.8 \times 10^{2}$$

This method of expressing numbers is called *scientific notation*.

The advantage of this notation is to help in keeping the decimal point straight in a long computation. Suppose we needed to find the answer to the following problem.

$$\frac{20,000 \times 5,000 \times 800}{400,000}$$

An easy way to do this is to use scientific notation:

$$20,000 = 2 \times 10^{4}$$
$$5,000 = 5 \times 10^{3}$$
$$800 = 8 \times 10^{2}$$
$$400,000 = 4 \times 10^{5}$$

The expression can then be changed to

$$\frac{2 \times 10^{4} \times 5 \times 10^{3} \times 8 \times 10^{2}}{4 \times 10^{5}}.$$

By rearranging the order of the factors, and then regrouping the factors, we obtain

$$\frac{2 \times 5 \times 8}{4} \times \frac{10^{4} \times 10^{3} \times 10^{2}}{10^{5}},$$

which simplifies to

$$20 \times 10^{4+3+2-5} \text{ or } 20 \times 10^{4} \text{ or } 200,000.$$

Now let's work an example that involves very small numbers.

$$\frac{.0002 \times .005 \times .8}{.000004}$$

Each number could be changed to scientific notation:

$$.0002 \ = 2 \times 10^{-4}$$
$$.005 \ = 5 \times 10^{-3}$$
$$.8 \ = 8 \times 10^{-1}$$
$$.000004 = 4 \times 10^{-6}.$$

If we put these back into the expression, we get

$$\frac{2 \times 10^{-4} \times 5 \times 10^{-3} \times 8 \times 10^{-1}}{4 \times 10^{-6}}$$

By rearranging and regrouping, we can write this as

$$\frac{2 \times 5 \times 8}{4} \times \frac{10^{-4} \times 10^{-3} \times 10^{-1}}{10^{-6}},$$

which simplifies to

$$20 \times 10^{-4-3-1-(-6)} = 20 \times 10^{-8+6} = 20 \times 10^{-2} \text{ or } .2.$$

EXERCISE SET 4
Using Scientific Notation

1. Write each of the following numbers as a product of a number between 1 and 10 and a power of 10.

    a. 24   b. 368   c. 4,270   d. .23   e. .0481   f. 6.42

2. Use scientific notation to perform the following computations.

    a. $80,000 \times 5,000$ b. $20,000 \times 16,000,000$

    c. $.0006 \times 8,000$ d. $.0006 \div 8,000$

    e. $\dfrac{4,000 \times 1,200,000}{240,000}$ f. $\dfrac{.002 \times .005 \times .01}{.00001}$

The Logarithm Story

If any number can be expressed as a power of 10, then the powers of 10 can be used in computations involving any numbers. We have seen that any number greater than 10 can be expressed as a product of a number between 1 and 10 and a power of 10. We now have the question, "Can any number between 1 and 10 be expressed as a power of 10?" We know that $1 = 10^0$ and $10 = 10^1$. Therefore, if any number between 1 and 10 is to be expressed as a power of 10, the exponent would have to have a value between 0 and 1. But there is still the problem of finding the different powers of 10 for the numbers between 1 and 10.

John Napier, with the help of an English geometry professor named Henry Briggs, solved this problem in the early part of the seventeenth century. A partial list of their results, containing the whole numbers between 1 and 10 expressed as powers of 10, is given in Table 4. The exponents in the table have been rounded off to the nearest thousandth.

Table 4

Power of 10	10^0	$10^{.301}$	$10^{.477}$	$10^{.602}$	$10^{.699}$	$10^{.778}$	$10^{.845}$	$10^{.903}$	$10^{.954}$	10^1
Number	1	2	3	4	5	6	7	8	9	10

Napier gave the name *logarithms* to his discovery. Napier's logarithms were exponents of the base 10. Today, any set of exponents used for computing is called a system of logarithms, and the exponents of the base 10 are called *common logarithms*.

A more complete list of common logarithms is given in Table 5. The numerals in the column marked N on the left side of the table represent whole numbers, while those across the top represent tenths. The numerals in the body of the table are the powers of 10 (to the nearest thousandth) of the numbers represented by the left column and top row. There should be a decimal point in front of each numeral in the body of the table, but the decimal points have been omitted to save space.

Table 5
Table of Three-Place Common Logarithms

N	0	1	2	3	4	5	6	7	8	9
1	000	041	079	114	146	176	204	230	255	279
2	301	322	342	362	380	398	415	431	447	462
3	477	491	505	519	531	544	556	568	580	591
4	602	613	623	633	643	653	663	672	681	690
5	699	708	716	724	732	740	748	756	763	771
6	778	785	792	799	806	813	820	826	833	839
7	845	851	857	863	869	875	881	886	892	898
8	903	908	914	919	924	929	934	939	944	949
9	954	959	964	968	973	978	982	987	991	996

We can use Table 5 to find the power to which we would have to raise 10 to get any two-digit number from 1.0 to 9.9. Suppose you wanted to find the logarithm of 3.4 (abbreviated as log 3.4). You go down the vertical column labeled N until you come to 3, then follow this third row over to the column with the heading of 4. The reading in the table is 531. You know, therefore, that

$$\log 3.4 = .531 \text{ or } 3.4 = 10^{.531}.$$

The methods used to compute the common logarithms of numbers between 1 and 10 are too complex to discuss here. However, we can learn how to use and understand the meaning of logarithms.

Since we can express any number greater than 10 as a product of a number between 1 and 10 and a power of 10 and can use Table 4 to express numbers between 1 and 10 as powers of 10, we can now write any number as a power of 10. Of course, this is limited by the extent of the logarithm table being used. To find the logarithm of any number, we would express the number in scientific notation, use a logarithm table to write the factor between 1 and 10 as a power of 10, then use the law for the multiplication of exponential quantities to obtain the complete logarithm. For example,

$\log 2 = .301$, because $2 = 10^{.301}$;

$\log 20 = 1.301$, because $20 = 2 \times 10 = 10^{.301} \times 10^1 = 10^{.301+1} = 10^{1.301}$;

$\log 200 = 2.301$, because $200 = 2 \times 100 = 10^{.301} \times 10^2 = 10^{.301+2} = 10^{2.301}$;

$\log 2,000 = 3.301$, because $2,000 = 2 \times 1,000 = 10^{.301} \times 10^3 = 10^{.301+3} = 10^{3.301}$

In the other direction, the logarithms of .2, .02, .002, and so on, are also closely related:

$\log .2 = .301 - 1$, because $.2 = 2 \times 10^{-1} = 10^{.301} \times 10^{-1} = 10^{.301-1}$;

$\log .02 = .301 - 2$, because $.02 = 2 \times 10^{-2} = 10^{.301} \times 10^{-2} = 10^{.301-2}$;

$\log .002 = .301 - 3$, because $.002 = 2 \times 10^{-3} = 10^{.301} \times 10^{-3} = 10^{.301-3}$.

Notice that .301 (which is log 2) appears in all of these logarithms. The amount added to or subtracted from .301 is the exponent of 10 used in expressing the number in scientific notation.

The decimal fraction part of a logarithm is called the *mantissa* of the logarithm of a number. The whole number to be added to or subtracted from the mantissa is called the *characteristic* of the logarithm of the number.

If we know the logarithm of a number, Table 5 can be used to find the number. Suppose we want to find a number (x) having a logarithm of 2.833. We can say that

$$\log x = 2.833$$
$$x = 10^{2.833}$$
$$x = 10^2 \times 10^{.833}$$
$$x = 100 \times 6.8$$
$$x = 680.$$

The 6.8 was obtained by looking up 833 in the body of Table 5 and finding the number that corresponded to it.

EXERCISE SET 5
Understanding Logarithms

1. Use Table 5 to find the logarithms of the following numbers.
   a. 82   b. 6   c. .07   d. 850   e. .49   f. .00012
2. Use Table 5 to find the numbers having the following logarithms.
   a. 1.716   b. 0.477   c. 3.415   d. .041 − 2   e. .982 − 1   f. − 5
3. If log 2 = .30103, log 3 = .47712, and log 10 = 1, find the logarithms of the following without the use of tables.

   a. log 5   (*Hint:* $5 = \dfrac{10}{2}$)
   b. log 6   (*Hint:* $6 = 2 \times 3$)
   c. log 8   (*Hint:* $8 = 2 \times 2 \times 2$)
   d. log 9   (*Hint:* $9 = 3 \times 3$)
   e. log 60   (*Hint:* $60 = 2 \times 3 \times 10$)

$$M = \frac{3.833 + .613}{2}$$
$$M = 10^{1.223}$$
$$M = \frac{\text{LOG } 6{,}800 + \text{LOG } .041}{2}$$

Computing with Logs

We are now ready to put our ideas about logarithms to work. Since common logarithms are exponents of the base 10, we can use them to do computations just as we used the exponents of the number 2 for computations.

The laws of exponents that were developed in the work with the base 2 can be applied to common logarithms. If we wish to multiply two numbers, we add their logarithms; if we wish to divide one number by another, we subtract the logarithm of the divisor from that of the dividend; if we wish to raise a number to a power, we multiply the logarithm of the number by the power; if we wish to find the root of a number, we divide the logarithm of the number by the index of the root.

The following examples illustrate the value of logarithms in doing computations.

Example 1.

Find the value of M if

$$M = \sqrt{6,800 \times .041}.$$

The solution can be organized as follows:

$$\log M = \frac{\log 6,800 + \log .041}{2}$$

$$\log M = \frac{3.833 + .613 - 2}{2}$$

$$\log M = \frac{3 - 2 + .833 + .613}{2}$$

$$\log M = \frac{1 + 1.446}{2}$$

$$\log M = \frac{2.446}{2} = 1.223$$

$$M = 10^{1.223}$$

$$M = 17 \text{ (approximately)}.$$

Example 2.

Find the value of M if

$$M = \frac{(23)^2}{77}.$$

The solution can be organized as follows:

$$\log M = 2(\log 23) - \log 77$$

$$\log M = 2(1.362) - 1.886$$

$$\log M = 2.724 - 1.886 = .838$$

$$M = 10^{.838}$$

$$M = 6.9 \text{ (approximately)}.$$

In both examples, the laws of exponents were used to set up the problem, the logarithms of the numbers were found in Table 5, the simple calculations were performed, and then Table 5 was used to find the number that corresponded to the logarithm of the answer.

We can see the great value of logarithms, for multiplications and divisions are reduced to additions and subtractions, and the job of finding powers and roots is changed to one of performing simple multiplications and divisions. Although common logarithms have 10 as a base, this is not a requirement for all logarithms. Logarithms for special purposes are figured on other bases.

John Napier's great discovery has been used by astronomers in charting the stars; by mariners in plotting their position on the seas; and by engineers in building cities, highways, and dams. Engineers and scientists still use logarithms, although modern computing machines have taken over some of the jobs formerly done by logarithms. Nevertheless, logarithms will continue to play an important role in the field of mathematics.

EXERCISE SET 6
Using Logarithms

1. Use Table 5 to find the value of M in the following.
   a. $M = \log 8$ b. $\log M = 8$ c. $M = \log 74$
   d. $\log M = .740$ e. $M = \log 29$ f. $\log M = .079 - 2$

2. Use logarithms to evaluate M if
   a. $M = \sqrt{82}$ b. $M = 2.3^4$
   c. $M = \dfrac{4{,}700 \times 85{,}000}{270{,}000}$ d. $M = \dfrac{.0037 \times .074}{.00016}$

Computing with Rulers

When you were learning addition and subtraction combinations in your early years in school, you probably wished for some type of device that would do the work for you. You could have provided such a device by getting a pair of yardsticks. Figure 14 shows how a pair of yardsticks could be used to add 5 and 7. One stick is placed above the other, and the upper stick is slid to the right until its left end is above the seven-inch mark on the other stick.

Then the five-inch mark is found on the upper stick, and the answer, 12, is read directly below it on the lower stick. In a reverse manner, the yardsticks can be used for subtractions. Thus, to compute 12 − 5, place the five-inch mark on the top stick directly over the twelve-inch mark on the lower stick. The answer, 7, is found on the lower stick under the left edge of the top stick.

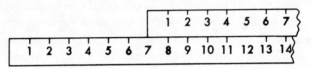

Figure 14

A pair of meter sticks marked in millimeters can be used to do additions and subtractions to three decimal places. Thus we see that a pair of marked sticks (often called a *slide rule*) can be used to do additions and subtractions mechanically. Can such a method be used to do multiplication and division? The answer is yes, for we know that, through the use of logarithms, multiplications can be done by performing additions, and divisions can be done by performing subtractions. However, the marks on the scales of a multiplication and division slide rule cannot be equally spaced, but must be based on values from a table of logarithms.

Shortly after John Napier discovered logarithms, William Oughtred (1574-1660), an English mathematician, recognized the fact that this discovery could be used to do multiplication and division mechanically. In 1622, he used two sticks marked with logarithmic scales to make a simple slide rule for multiplication and division.

In order to understand the construction of a slide rule for multiplication and division, let's use Table 6, which gives the common logarithms of the whole numbers from 1 to 10.

Table 6

Number	1	2	3	4	5	6	7	8	9	10
Logarithm	0	.301	.477	.602	.699	.778	.845	.903	.954	1

The multiplication of 2 by 3 can be done by adding the logarithms .301 and .477 and finding the number, 6, that corresponds to the sum of the logs (log 6 = .778 = .301 + .477). This addition can be done mechanically by using a pair of scales, each marked in the manner illustrated in Figure 15.

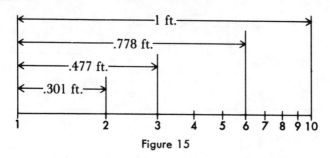

Figure 15

Marks of 1, 2, 3, 4, 5, 6, 7, 8, 9, and 10 are located on the scale according to the value of their logarithms — that is, 1 is at the left end, because log 1 = 0; 2 is marked at 301 thousandths of the length of the scale, because log 2 = .301; 3 is marked at 477 thousandths of the length of the scale, because log 3 = .477; and so on, finishing with 10 at the right end, because log 10 = 1. *The various lengths on the scale represent the logarithms of the numbers from 1 to 10.*

When you use two of these scales side by side, you can easily add logarithms and read the answer as the product of the two numbers. In order to set numbers accurately to do the multiplications and divisions, commercial slide rules are constructed with two movable parts — the sliding scale and the runner with a cross wire or hairline. Figure 16 pictures a modern slide rule. Notice that there are many lengths marked on the scale to represent the logarithms of decimal fraction values between 1 and 10.

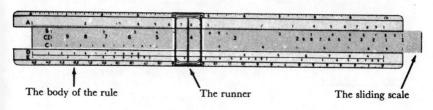

The body of the rule   The runner   The sliding scale

Figure 16

The two principal slide rule scales that are used for multiplication and division are usually called the *C* and *D* scales. To multiply two numbers, such as 2 × 3, you set the hairline on the first number, 2, on the *D* scale. Move the *C* scale so that the 1 (called the left index) falls on the hairline. Then move the hairline to the second number, 3, on the *C* scale. The product appears on the *D* scale at 6, directly below 3 on the *C* scale. (See Figure 17 on page 264.)

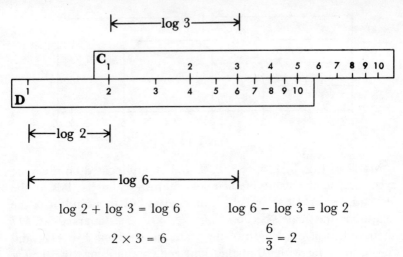

$$\log 2 + \log 3 = \log 6 \qquad \log 6 - \log 3 = \log 2$$
$$2 \times 3 = 6 \qquad \frac{6}{3} = 2$$

Figure 17

For division, such as $\frac{6}{3}$, you reverse the process. The numerator, 6, is located on the D scale by means of the hairline. Move the C scale until the divisor, 3, appears at the hairline above 6 on the D scale. The answer, 2, appears below the left index of the C scale.

Figure 17 shows how the slide rule finds the product of 2×3 by mechanically adding logarithms, and the quotient $\frac{6}{3}$ by mechanically subtracting logarithms.

If we multiply 4 times 5, placing the left index of the C scale on 5 of the D scale, we find that 4 has gone past the right end of the D scale.

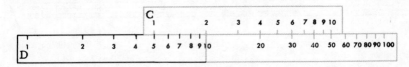

Figure 18

Imagine a second D scale placed so that its left index coincides with the right index of the regular D scale, as illustrated in Figure 18. The two scales together could be thought of as representing numbers from 1 to 100. On the second D scale, under 4 of the C scale, we read 20, the product of 4 and 5. We obtain this result because $\log 4 + \log 5 = \log 10 + \log 2 = \log 20$. A slide rule could be made with several C and D scales connected end to end so that an answer would never come off the end of a scale, but this is not practical. Furthermore, it is not necessary, for the product of 4 and 5 can be read by placing the right index of the *C* scale over 5 of the *D* scale. We can then read 2 on the *D* scale under 4 of the *C* scale as shown in Figure 19.

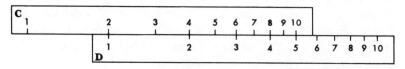

Figure 19

In Figure 19 the *C* and *D* scales are in the same relative positions as the *C* scale and the imaginary extension of the *D* scale shown in Figure 18.

Figure 20 shows the relationship between Figures 18 and 19. By eliminating an extension of the *D* scale, we are really subtracting one scale length. Since these are logarithmic scales, subtracting one scale length is equivalent to dividing by 10. Therefore, when we use only two scales to multiply 4 by 5 and obtain the result 2, we must realize that the result is really 2×10.

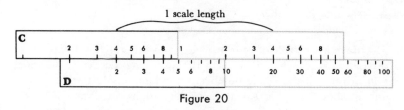

Figure 20

Any time we use the slide rule for computations, we have the problem of determining the placement of the decimal point. If we were to do the following multiplication by slide rule —

$$674 \times 75.4$$

— we would read the digits, 508, from the slide rule as the answer. The decimal point can be located by writing each number in the product in scientific notation:

$$674 \times 75.4 = 6.74 \times 10^2 \times 7.54 \times 10,$$

or approximately

$$7 \times 7 \times 10^2 \times 10,$$

which equals

$$49 \times 10^{2+1} = 49 \times 10^3.$$

Therefore, the answer would be 50,800, to three digits of accuracy. In general, this method for estimating the placement of the decimal point can be done in your head.

As you can see, the slide rule is a very useful computing device. Its chief advantage is that it can enable you to give rapidly a good estimate of the answer to a difficult computation. Its chief disadvantage is that it can only give answers of three or four figures. Engineers and scientists would use at least four- or five-place logarithm tables or calculating machines if they needed greater accuracy than that of the slide rule.

If you purchase a slide rule, you will find other useful scales that can do special things for you, such as squaring or cubing numbers or finding square roots and cube roots. Some slide rules have scales that solve problems in trigonometry, and there are many other scales that can be used to solve even more special problems. Slide rules that you can buy have instructions with them, and, with practice, you can solve many kinds of problems.

EXERCISE SET 7
Making Slide Rules

1. Make a slide rule of your own. Use two strips of graph paper with 10 divisions to the inch and a total of 10 inches long. Locate on the scales the logarithms of the whole numbers from 1 to 10. Write these whole numbers above the marks where the logarithms are located. Locate tenths also. Use Table 5 to obtain the necessary values. Mount the scales on cardboard or wood and you can use them for multiplying and dividing.

2. You can also make a slide rule by buying semi-log graph paper at a stationery store. In this case, all you have to do is cut two strips, mount them, and label each scale properly.

3. You can make a giant slide rule, too, if you wish, by using two meter sticks, relabeling each to correspond to the logarithms of numbers from 1 to 10.

4. Suppose that you solved this problem by slide rule —

$$\frac{234 \times 671}{35.6}$$

— and found the digits of the answer to be 441. Locate the decimal point by using scientific notation.

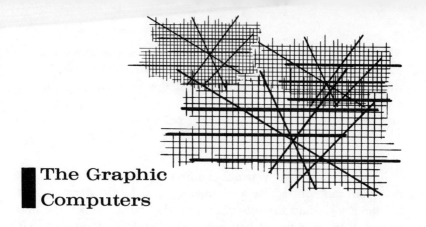

The Graphic Computers

Adding and Subtracting with a Nomograph

You have often worked with graphs in your study of mathematics, but have you ever used a graph for computing? Devices called *nomographs* have been designed to allow you to do rapid computing by just reading numbers from scales drawn on notebook paper or graph paper. One type of nomograph is shown in Figure 21. To make such a nomograph, start with three equally spaced, horizontal, parallel lines, *A*, *B*, *C*. Draw a vertical line, *D*, crossing each of the three lines at a point to be designated as the origin or zero point. Mark off equal spaces on each line. Starting at the zero points on lines *A* and *C*, number corresponding marks uniformly with signed numbers. Mark each point on line *B* with a numerical value twice that of the corresponding marks on lines *A* and *C*.

By placing a straightedge across the three lines, you will obtain a reading on each scale. These readings are related to one another by the formula

$$A + C = B.$$

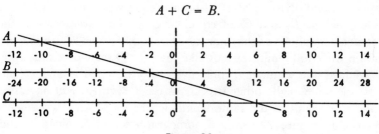

Figure 21

If you wish to add any two signed numbers, you locate one on scale A and the other on scale C. The line joining these two points will cross B at a point that represents the answer. For example, Figure 21 illustrates the solution

$$(- 10) + (6) = (- 4).$$
$$\uparrow \qquad \uparrow \qquad \uparrow$$
$$A \quad + \quad C \quad = \quad B$$

If $A + C = B$, then you also know that $C = B - A$. This says that you can subtract two numbers if you locate the minuend on the B scale and the subtrahend on the A scale. Then the line joining these two points will cross the C scale at the point that represents the difference, $B - A$. Figure 21 illustrates the problem.

$$(6) = (- 4) - (- 10).$$
$$\uparrow \qquad \uparrow \qquad \uparrow$$
$$C \quad = \quad B \quad - \quad A$$

EXERCISE SET 8
Using the Addition-Subtraction Nomograph

Make a nomograph like the one in Figure 21 and use it to verify the following additions and subtractions of signed numbers.

1.	$(+ 2) + (+ 4)$	**5.**	$(+ 2) - (+ 4)$
2.	$(+ 2) + (- 4)$	**6.**	$(+ 2) - (- 4)$
3.	$(- 2) + (+ 4)$	**7.**	$(- 2) - (+ 4)$
4.	$(- 2) + (- 4)$	**8.**	$(- 2) - (- 4)$

Multiplying and Dividing with a Nomograph

It is possible to make a nomograph that can be used for multiplication and division. Start with three equally spaced, vertical line segments, A, B, and C, of equal length. The lower ends of the segments should be on the same horizontal line. A and C are marked off on their full length exactly like the C or D scale of a slide rule. The third line segment B, placed halfway between A and C, has half the scale of the other two. Thus the B segment will have two slide rule scales in the same length where the A and C segments have one. Figure 22 shows a nomograph for multiplication and division.

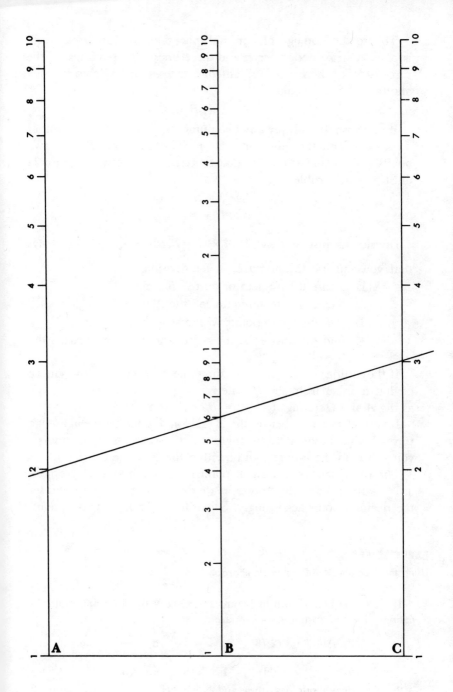

Figure 22

To use the nomograph for multiplication, all you need to do is place a straightedge across the three lines. The readings on the three scales where the straightedge crosses are related to one another by the formula

$$A \times C = B.$$

If you wish to multiply any two numbers, you locate one number on scale A and the other on scale C. The line joining these two points will cross B at a point that represents the answer. Figure 22 illustrates the problem.

$$2 \times 3 = 6.$$
$$A \times C = B$$

Division is just as easy. If $A \times C = B$, then $\dfrac{B}{A} = C$. This says that you can use the nomograph for division if you

(1) locate the numerator on the B scale,
(2) locate the denominator on the A scale,
(3) join these two points with a straight line, and
(4) find the answer at the point where the line crosses the C scale.

If the settings on A and C are the same, you will find the square of that number from the B scale reading that is on the same line as the A and C settings ($A \times A = B$ or $A^2 = B$).

If you select a number on the B scale and place the straightedge through that point at right angles to the segments A, B, and C, you will read the square root on either the A or the C scale.

The placement of a decimal point is the same kind of problem as it is with a slide rule. You estimate the size of the answer and set the decimal point accordingly. Scientific notation is again useful.

EXERCISE SET 9
Using the Multiplication-Division Nomograph

1. Use the nomograph in Figure 22 to verify the following multiplications, divisions, squares, and square roots.

a. $(2)(4)$ b. $(2)(8)$ c. $\dfrac{24}{3}$ d. $\dfrac{.0032}{.016}$ e. $(1.5)^2$

f. $(5)^2$ g. $\sqrt{9}$ h. $\sqrt{16}$ i. $\sqrt{5}$ j. $\sqrt{50}$

2. Use ideas about logarithms and exponents to explain why Figure 22 can be used to compute squares and square roots.

Computing with Machines

The History of Calculating Machines

In 1642, a nineteen-year-old French lad named Blaise Pascal was working in his father's tax office in Rouen, France. He became so irritated with the task of adding long columns of figures that he built a mechanical calculating machine. Young Pascal based his invention on gear wheels. There was a wheel with ten notches for each place value up to a hundred thousand. When a wheel had moved ten notches after one revolution, it was geared to make the next wheel to the left move one notch. The wheels were moved by dialing numbers in the various place-value columns.

Figure 23

A machine that could multiply automatically was the next major improvement. Such a machine was built by Gottfried Wilhelm von Leibniz (1646-1716), the German mathematician who discovered the principles of calculus at about the same time as Sir Isaac Newton.

In 1671, Leibniz developed a machine that could do multiplication by rapidly repeated addition. Figure-wheels on the fixed part of the machine registered the results of repeatedly adding up to nine times the value of each digit in the multiplicand. The multiplicand was represented on a sliding portion of the machine that could be moved by steps into positions corresponding to units, tens, hundreds, and so on. An important part that is still used in some machines today was a cylindrical drum, called a "stepped wheel," having on a part of its outer surface nine teeth of increasing length from one to nine units.

As time went on, the ideas of Pascal and Leibniz were improved and expanded. New developments eventually led to the desk calculators of the present day.

In 1812, a new idea in computers came to life. An Englishman named Charles Babbage started to develop ideas for a steam-driven machine that was to compute and print mathematical tables. He planned a machine that would not depend upon a human being to push a key or turn a crank for each calculation step, but instead would do the calculation automatically. By 1882, Babbage had developed a small working model of such a machine, but he was never able to finish a full-scale model. He also planned a larger automatic calculator that was to be controlled by the use of holes punched in cards, but, again, he was unable to complete it. The factories of his time could not turn out the precision gears and cogs needed for the machines. Although Charles Babbage died a disappointed man, his designs for calculators lived on and became the basis for many of the ideas used in modern electronic computers.

A second major contribution to modern calculators was made in the nineteenth century. It was discovered that the height of the tide at seaports could be computed from a complex formula related to the motions of the sun and the moon. In 1872, the Englishman Lord Kelvin built a machine made up of pulleys, weights, and connecting cords that was to be used to predict the height of the tides. When operated, the machine gave a physical measurement that was proportional to the result for the calculation of the tide at a certain time. Hence a measurement from the machine was used to give a mathematical computation.

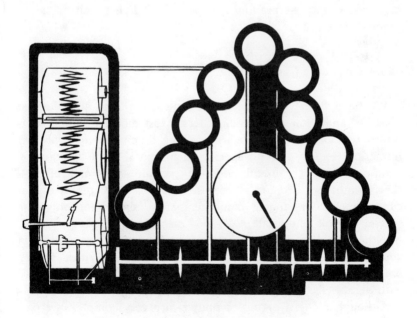

Figure 24

Many of the automatic computing machines of today have

resulted from either the basic ideas of Charles Babbage or those of Lord Kelvin. The modern electronic computers are classified as either *digital* or *analog* computers. A digital computer, such as the one proposed by Babbage, carries out computations by counting or in some way working with the digits of the original problem. An analog computer, like that invented by Lord Kelvin, turns out answers through the use of physical measurements, such as length, weight, or voltage, that are analogous (proportional) to the values in the desired calculation. An analog computer measures one type of quantity, but gives a result in terms of some other quantity.

The terms "digital" and "analog" can be applied to devices other than the modern electronic computers. An abacus, a cash register, and an ordinary adding machine are examples of digital computers. A slide rule, an automobile speedometer, and a thermometer are examples of analog computers. A slide rule uses lengths to compute any type of numerical answer; an automobile speedometer counts turns of the wheel, but reads miles per hour; a thermometer reading is dependent upon the height of a column of mercury, but the reading is expressed in Fahrenheit or centigrade degrees.

Modern Electronic Computers

We have seen that modern electronic computers are classified as either analog or digital computers. Most modern analog computers use variations in electric voltage or current to represent numerical values and do special calculations. Although analog computers are very useful in solving certain types of problems, they cannot be used for as wide a range of problems as digital computers. Often analog computers are designed for only one special type of computation. Analog calculators compute rapidly, but they are less accurate than electronic digital computers. As you might guess, digital electronic computers are more widely used than the analog type. Therefore, we will be concerned mainly with digital electronic computers.

Although modern digital computers are very complex, we can get some general ideas about the basic principles involved in their construction and operation. An electronic digital computer performs calculations in much the same way as you would do an arithmetic problem on an abacus. You would use the abacus for counting; a piece of paper to record the problem, results, and any facts needed to work the problem; and your own mind to control the entire operation. It would also be necessary for you to have learned the steps required to do the calculation.

A digital computer has an *arithmetic unit*, corresponding to your abacus, that does the counting and calculating necessary to solve a problem. It has a *storage* or *memory unit*, corresponding to your paper, in which the numbers and other facts needed to work the problem are stored. It has a *control unit*, corresponding to your mind, that makes sure the steps in the calculation are carried out in the proper order. But how does the machine know the proper steps for a computation? The steps needed to solve a problem must be furnished by the person using the computer. He must supply the data and instructions for handling the data.

Numerical data or instructions represented by a numerical code are usually put into a computer by using punched cards or magnetic tape. This information is then transferred from the cards or the tape to the storage part of the computer. The control section sees that the instructions are carried out, the actual computation being done in the arithmetic section. Results are transferred back to the storage section and can be taken out of the machine in the form of punched cards or magnetized tape, which then can be interpreted. Figure 25 shows a simple operational scheme for a digital computer.

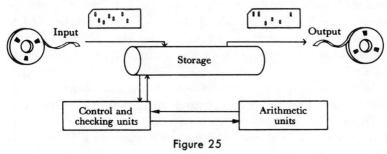

Figure 25

An electronic digital computer uses magnetism and electricity to represent numbers and perform arithmetic operations.

The storage section of a computer consists of a rotating magnetic drum. The drum is divided into thousands of small sections, and each section can represent a number. If a spot in the section is magnetized in one way (perhaps as a positive pole), it represents a 1; if it is magnetized in the other direction (perhaps as a negative pole), it represents a 0.

A computer also contains thousands of simple electronic circuits (called relays) and hundreds of vacuum tubes or transistors (like the tubes and transistors in a radio). These devices are used to represent numbers, transfer information from one part of the machine to another, and control the operation of the machine. They also enable the machine to count and compute. If an electric pulse is sent through a circuit or a tube, a count of 1 is registered; if no pulse is sent through the circuit, the count is 0. Figure 26 shows how a relay can count pulses sent through it by using the position of a switch to represent a 1 or a 0.

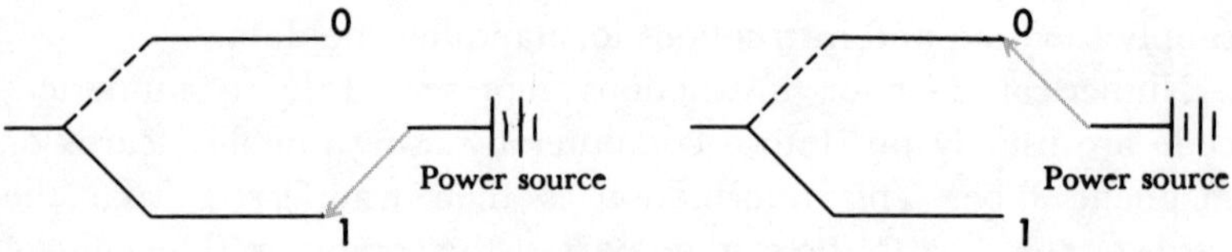

Pulse sent through circuit — switch in 1 position   No pulse — switch in 0 position

Figure 26

The behavior of the magnetic drum, relays, and vacuum tubes suggests that the counting process in an electronic computer should be based on a *binary* numeration system that involves only the digits 0 and 1. Combinations of these two digits can be used to represent any number. A comparison between counting in the binary system and counting in the decimal system is shown in Table 7.

In the "ten" system, carrying occurs when you count to 9 and add 1 more. Thus $9 + 1$ is registered as 10, meaning 1 "ten" and 0 "ones." In the binary (two) system, you start by counting: 1 and 1 more make a group of "two"; this is written as 10, meaning 1 "two" and 0 "ones."

The system of numeration described in Table 7 is based on powers of 2 and is sometimes called a *base two* system. Numerals written in this system have the base indicated to the right and below the numeral (a subscript). For example, 1001_{two} in the binary system and the symbol 9 in the decimal system represent the same quantity.

Table 7

	Counting in the Binary System						Counting in the Decimal System		
2^5	2^4	2^3	2^2	2^1	2^0	10^2	10^1	10^0
$(2\times2\times2\times2\times2)$	$(2\times2\times2\times2)$	$(2\times2\times2)$	(2×2)	(2×1)	1	10×10	10×1	1
32's	16's	8's	4's	2's	1's	100's	10's	1's
					1			1
				1	0			2
				1	1			3
			1	0	0			4
			1	0	1			5
			1	1	0			6
			1	1	1			7
		1	0	0	0			8
		1	0	0	1			9
		1	0	1	0		1	0
		1	0	1	1		1	1
		1	1	0	0		1	2
		1	1	0	1		1	3
		1	1	1	0		1	4
		1	1	1	1		1	5
	1	0	0	0	0		1	6
	1	0	0	0	1		1	7
	1	0	0	1	0		1	8
	1	0	0	1	1		1	9
	1	0	1	0	0		2	0
	1	0	1	0	1		2	1
	1	0	1	1	0		2	2
	1	0	1	1	1		2	3
	1	1	0	0	0		2	4
	1	1	0	0	1		2	5
	1	1	0	1	0		2	6
	1	1	0	1	1		2	7
	1	1	1	0	0		2	8
	1	1	1	0	1		2	9
	1	1	1	1	0		3	0
	1	1	1	1	1		3	1
1	0	0	0	0	0		3	2

The binary addition table shows when the "carry" occurs in base two counting.

$$\begin{array}{c|cc} + & 0 & 1 \\ \hline 0 & 0 & 1 \\ 1 & 1 & 0 + \text{"carry" of 1 or } 10_{two} \end{array}$$

A "carry" is generated whenever a 1 is to be added to 1 already present. Any digit receiving a 1 as a result of "carrying" is complemented, that is, a 0 becomes 1 and a 1 becomes 0.

You can get a better feeling of how an electronic computer must operate by actually adding numbers expressed in the binary notation.

Add:

$$\begin{array}{r} 1\ 0\ 1_{two} \\ +\quad 1 \\ \hline 1\ 1\ 0_{two} \end{array}$$

Perform steps from right to left.
$(1 + 1 = 0 + \text{"carry"})$
$(0 + \text{"carry"} = 1)$

$$\begin{array}{r} 1\ 1\ 0\ 1_{two} \\ +\ 1\ 1\ 0\ 0_{two} \\ \hline 1\ 1\ 0\ 0\ 1_{two} \end{array}$$

Perform steps from right to left.

(1 from "carry")
$(1 + 1 = 0 + \text{"carry"})$
$(1 + 0 = 1)$
$(1 + 1 + 1 = 1 + \text{"carry"})$
$(0 + 0 = 0)$

You can see by Table 7 that it takes many places in the binary system just to represent very small numbers. To avoid this difficulty, some electronic computers use a binary-decimal system, that is, a combination of the decimal and binary systems. The decimal system is used in the usual form, but each digit of a decimal numeral is put in the binary form. Thus, instead of expressing an entire number like

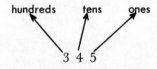

hundreds   tens   ones

3 4 5

in the binary form, each digit, the 3, the 4, and the 5, is expressed in the binary form, with different sets of relays used for ones, hundreds, and thousands. Thus 3 becomes 11 in the binary system, 4 becomes 100 in the binary system, and 5 becomes 101 in the binary system.

hundreds	tens	ones
11	100	101
3	4	5

The principle of computing by electronic devices is one of counting electric pulses and arranging electric circuits to represent a count so that, when a "place" is filled, the switching arrangements will "carry" to the next place. Let's try to put our ideas about binary numbers, electric pulses, and electric circuits together to see how a modern digital computer does arithmetic.

Suppose we have a series of relays which are capable of switching between two positions. We could consider one position as 0 and the other as 1. If a steady current is passed through the relays, the switches will not change position, but when a sudden pulse of current is sent through, the condition of the relay is changed. If the relay is in condition 0 when it receives a pulse, it changes to condition 1, but does not transmit a pulse to the next relay. It will only transmit a steady flow of current. If the relay is in condition 1 when it receives a pulse, it transmits the pulse to the next relay and changes its condition to 0.

Figure 27 shows a series of relays with all the switches at the 0 setting.

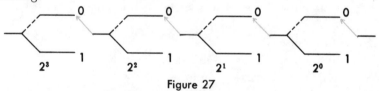

Figure 27

If a pulse, representing a count of 1, enters at the first relay, it will cause the first switch to change to the 1 position, as shown in Figure 28.

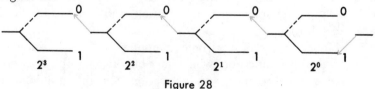

Figure 28

A second pulse will cause the second switch to move to the 1 position, but will also cause the first switch to move back to the

0 position. Thus a count of 2 will be represented by the relays as shown in Figure 29.

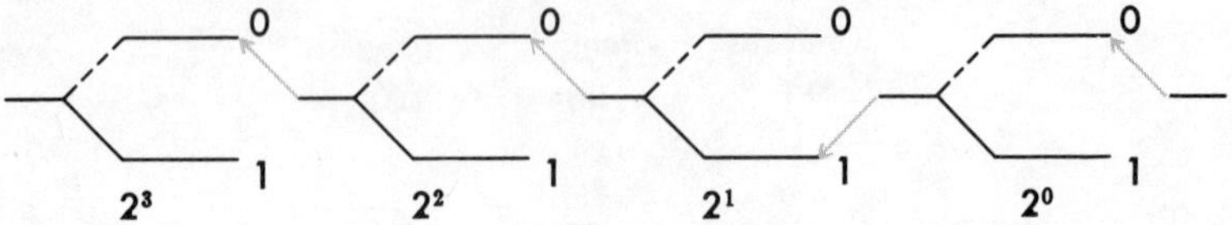

Figure 29

A third pulse will merely cause the first switch to move back to the 1 position. This process continues in exactly the same way as each new pulse enters the series of relays.

If you write down the settings of the switches at each pulse, you can see the binary system in action:

Initial positions are all zeroed:	0 0 0 0
1st pulse:	0 0 0 1
2nd pulse: (The 2nd switch is brought into play.)	0 0 1 0
3rd pulse:	0 0 1 1
4th pulse: (The 3rd switch is brought into play.)	0 1 0 0
5th pulse:	0 1 0 1
6th pulse:	0 1 1 0
7th pulse:	0 1 1 1
8th pulse: (The 4th switch is brought into play.)	1 0 0 0
9th pulse:	1 0 0 1
10th pulse: Zeros all switches and "carries" into a new set of switches to represent the "tens" place.	

In this way, an electronic computer can perform additions and subtractions by using an electronic binary counter. It can perform multiplications by doing repeated additions and divisions by doing repeated subtractions.

The binary nature of electronic computers makes it possible to use them to answer questions which have only two possible answers. If a card is punched in a certain location it might mean that a person identified by the card meets a given condition. Another card which is not punched would mean that the person identified by the card does not meet the condition. The computer can answer the question, "Is this the card of a person who meets the given condition?"

For example, you might punch cards to identify people according to their sex, color of hair, height, age, weight, marriage status and income. You could send a batch of cards to the machine, and, provided the cards are punched right and the machine is programmed right, you could have the machine sort out all blonde, unmarried men, 6 feet 2 inches in height, 21 years of age, weighing 170 pounds, and with an income of $5,000 a year or more.

The machine appears to be making decisions — this card passes and this one doesn't, this card is punched and this one isn't, this card completes the electric circuit because it is punched and this one doesn't because it is not punched. These machines are referred to as "electronic brains," but they really don't do their own thinking. Someone has to provide the machine with a basis on which it can make its decisions. In short, the electronic computers are no better than the mathematicians who use them. Any time a digital computer is used, whether it be for computations or simple decision making, it has to be told what to do. A mathematician sets up a problem by writing a "program" of instructions for the computer, and then the machine does the work in a very rapid fashion. The precision and speed of modern automatic computers is amazing. For example, in 1955, the value of π was computed to 3,089 places on a computer. This calculation took only thirteen minutes.

These modern developments are having a tremendous impact upon our civilization. Mathematical logic as a branch of mathematics is finding many practical applications dealing with the design, control, and use of these machines, particularly with respect to decision processes. The subject of numerical analysis has taken on new importance. People are finding that computing methods used on desk computers are not always applicable to the new machines. New methods of attack must be developed. New applications for the use of these machines are being found regularly. Problems that heretofore were beyond the powers of human computers are now being solved regularly.

These machines are being used for more and more applications in the fields of social science and business. This means that people in these areas will need to have a better understanding of mathematics than ever before if they are to make full use of what is available to them. Also, these machines have opened up many employment opportunities for those trained in mathematics. Modern computers need mathematicians to program problems for the machines. High interest and ability in mathematics are essential ingredients for success in this field, which you may find well worth your while to investigate.

EXERCISE SET 10
Questions on Computers

1. Classify each of the following as an analog or digital computer.
   - a. a clock
   - b. a counting board
   - c. your fingers
   - d. a bathroom scale

2. Use a relay diagram like the one in Figure 27 to represent the number 7.

3. Do the following base two additions.

   a. 1011_{two}
   $+ 1001_{two}$

   b. 11001_{two}
   $+ 111_{two}$

   c. 11111_{two}
   $+ 11111_{two}$

A Backward Glance and a Look to the Future

The story of computers has been a story of man's desire to free himself from the labors of tedious computations in order to have time to attack more important problems. Perhaps no more effective summary to this section can be given than an identification in chart form of the dates of mathematical events in history related to the development of computing devices.

In this chart you can see a picture of the mathematical progress of man. The most important impression the chart should leave with you is that we are really just on the threshold of many new advances. The so-called "electronic brains" are just in their infancy. Wouldn't you like to know what they will be like when they are fully grown? The most exciting prospect for the future is the fact that you will have an opportunity to play a part in its development. Perhaps you will build or operate a computer that will do calculations leading to the invention of a new plastic, the discovery a new rocket fuel, or the charting of the first trip to Mars.

A Time Chart of Computing Device Developments

9th Century B.C.	Place Value concepts were developed (Hindu-Arabic origin).
542 B.C.	Bamboo rods were used in China for calculating.
300 B.C.	Euclid developed geometry as a logical subject (Greece).
650 A.D.	First example of Hindu numerals used outside of India was found in Mesopotamia.
12th Century	The Chinese used an abacus made from beads and rods.
1614	Logarithms were invented by John Napier (Scotland).
1617	John Napier developed his numbering rods or "bones" to multiply easily (Scotland).
1620	Edmund Gunter built the first slide rule (England).
1624	The first comprehensive table of logarithms was published by Henry Briggs (England).

1642	The first calculating machine for adding and subtracting was invented by Blaise Pascal (France).
1671	The first calculating machine to do multiplications was invented by Leibniz (Germany).
1812	Charles Babbage developed many of the basic ideas used in modern calculators. However, he moved from one idea to another without ever completing a satisfactory working model (England).
1820	The first successful calculating machine manufactured on a commercial scale was invented by Charles Thomas of Colmar in Alsace, France.
1850	The runner was first added to the slide rule by Taverier-Gravet, a French company.
1850	The first key-driven adding machine was patented in the United States by D. D. Parmalee. It could add only a single column of digits at a time.
1872	Kelvin built his tide analyzer (England).
1876	The first tape adding machine was developed by E. D. Barbour of the U. S.
1883	The first cash register was developed by Patterson Brothers in Dayton, Ohio.
1887	The Comptometer, developed in the United States, was the first successful key-driven calculating machine that could handle more than one column of digits at a time.
1889	The invention of the punched card by Herman Hollerith of the United States laid the foundation for automatic sorting and tabulating machines.
1903	The "Duplex" model of the Comptometer was the first machine that could have simultaneous depression of keys in every column without interfering with the carrying of "tens."

1919	The first electronic "flip-flop" circuit that pertained to digital computers was described by W. H. Eccles and F. W. Jordan in a 1919 issue, of *Radio Review* (United States).
1931	Vannevar Bush of the Massachusetts Institute of Technology built the first "modern" large analog computer.
1931	C. E. Wynn-Williams published the first article about the use of thyratron tubes in counting circuits. It appeared in the *Proceedings of the Royal Society* (England).
1946	Projects of the United States Government in World War II came to light publicly for the first time by the appearance of the description of the electronic computer ENIAC in *Mathematical Tables and Other Aids to Computation*.
	From 1946 on, research has progressed rapidly throughout the world to develop greater capacity, greater speed, smaller size, and more applications for all types of analog and digital computers.

Extending Your Knowledge

If you would like to read more about computing devices, the following books can provide such information.

BANKS, J. HOUSTON, *Elements of Mathematics.* Allyn and Bacon, 1956

BERKELEY, E. C., *Giant Brains or Machines That Think.* John Wiley and Sons, 1949

CULBERTSON, J. T., *Mathematics and Logic for Digital Devices.* D. Van Nostrand Co., 1958

METZGER, R., *Elementary Mathematical Programming.* John Wiley and Sons, 1958

REINFELD, N., *Mathematical Programming.* Prentice-Hall, 1958

RICHARDS, R. K., *Arithmetic Operations in Digital Computers.* D. Van Nostrand Co., 1955

STIBLITZ, G., *Mathematics and Computers.* McGraw-Hill Book Co., 1957

SWAIN, ROBERT L., *Understanding Arithmetic*. Rinehart and Co., 1952

TOMPKINS, CHARLES B., "Computing Machines and Automatic Decisions," *Insights into Modern Mathematics*. Twenty-Third Yearbook, The National Council of Teachers of Mathematics, 1957

TURING, A. M., "Can a Machine Think?" in James Newman's *The World of Mathematics*. Simon and Schuster, 1956

NEUMANN, JOHN VON, "The General and Logical Theory of Automata," in James Newman's *The World of Mathematics*. Simon and Schuster, 1956

PART VI

THE
World
of Statistics

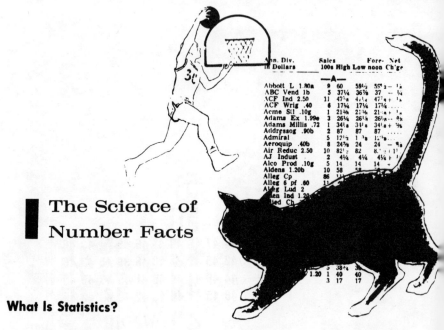

Ann. Div. in Dollars	Sales 100s	High	Low	Fore- noon	Net Ch'ge
—A—					
Abbott L 1.80a	9	60	59½	55⅞	2 — ⅛
ABC Vend 1b	5	37¼	36⅞	37	— ¼
ACF Ind 2.50	11	47⅝	4.¼	47⅞	+ ⅛
ACF Wrig .40	6	17¼	17¼	17¼	....
Acme Stl .10g	1	21¾	21¾	21¾	+ ⅛
Adams Ex 1.99e	3	26¼	26⅝	26⅝	— ⅝
Adams Millis .72	1	34⅛	34⅛	34⅛	+ ⅜
Addressog .90b	2	87	87	87	......
Admiral	5	12½	1⅜	1⅜	..
Aeroquip .40b	8	24⅝	24	24	— ⅝
Air Reduc 2.50	10	82½	82	8.¼	1¼
AJ Indust	2	4¼	4¼	4¼	+ ⅛
Alco Prod .10g	5	14	14	14	+
Aldens 1.20b	10	58	58		
Alleg Cp	86				
Alleg 6 pf .60	11				
Alleg Lud 2					
Allied Ch					
1.20	5	38¾	3.		
	1	40	40		
	3	17	17		

The Science of Number Facts

What Is Statistics?

"Curiosity once killed a cat," is an old saying. Like the cat, we are all curious about our world. What is going to happen tomorrow? What causes a tornado? How can a space ship go to Mars? What should I do to make friends? How long will I live? How can I get rich? Who will win the game next Friday? Will our curiosity about atomic energy cause the human race to become extinct?

Not all of these questions can be answered today. But the way to get answers to most questions is to collect facts about them. The facts we collect are often number facts like the batting averages of our favorite players. Number facts are called *data*, and the study of data is called *statistics*. Sometimes the number facts themselves are called statistics. However, it is better to think of statistics as a science which studies the collection and meaning of data. Many problems can be solved by statistics. The answers to problems that can be solved by statistics are often based on uncertainties and incomplete information. Such answers are only *probably* true, and are sometimes called predictions. A person who works with data for a living is called a statistician. A statistician must know a great deal about mathematics. He uses mathematics to increase the accuracy of his conclusions.

This part of the book will tell you about some of the ways you can collect and use data to discover new facts and ideas.

Using Statistics Today and Tomorrow

The world we live in today is a world that is changing rapidly. If you read the newspaper or listen to people talk, you will find that numbers are very often used in reporting the news. Questions concerning "how many," "how fast," "what time," "how much," and "where" usually need numbers for answers. So if you learn about the science of statistics, you will understand better what is happening in the world. Let's look at just a few ways in which statistics is being used in our world.

If you are interested in sports such as basketball, you probably read the statistics of the games. What does a player's shooting average mean? You may read that Mike's shooting average is .450, while Bill's is .350. This means that Mike has been making about 45 out of 100 shots and Bill about 35 out of 100. Does this mean that, if Mike and Bill take the same number of shots in a certain game, Mike will make more baskets than Bill? Maybe not. On a certain day, Bill might make more shots than Mike. On another day, Mike might make more than his average of 9 baskets out of 20 shots, while Bill might make less than 7 out of 20 shots. This is an example of statistics used in sports.

No matter where you look, you find statistics being used. For example, everybody talks about the weather. Weather talk is usually about the temperature, the wind, the rain, or the sunshine. Such information is recorded numerically at weather stations throughout the world. When the weatherman makes his forecast, he makes it only after he has studied the weather data that have been collected over a large region. He knows that he is taking a risk of being wrong in his forecast. This is an example of the use of numerical data in making predictions.

We are all interested in earning enough money to support ourselves. Information about business conditions, prices, wages, and unemployment is found in newspapers and magazines. This information helps people decide when to buy a new car, build a new house, ship livestock to the market, or buy government bonds. Businessmen watch these reports very closely to decide how much stock to have on their shelves, when to build a new factory, or when to plan a sale. The government also watches these reports very carefully to know how much tax money can be expected next year, how much it must pay for the goods it must buy, and how many miles of roads it must plan to build. Educators must study data to decide when and where to build new schools, how many teachers to hire next year, and what the school tax rate should be. In the same way, churches learn from statistics about changes in population, church attendance, and costs, so that new programs and new buildings can be planned. These examples point out the use of statistics in making decisions.

In many ways, statistics is used to change our world. Doctors study data from experiments to develop new medicines and to test their effectiveness. Businesses build laboratories and hire mathematicians and scientists to obtain information about new products such as building materials, home chemicals, automobile engines, rocket fuels, and high-speed computers. Electronic computers, in turn, make possible many new and important uses for statistics. Thus we see the use of statistics in discovering and testing new products, methods, and ideas.

Although statistics is a modern science, it is not entirely new. For example, in the Bible, the book of Numbers is full of numerical data. Since statistics was often used to study social problems in nineteenth-century England, it was called political arithmetic. Today statistics is used in almost all fields and is important throughout the world. One agency of the Federal government, the Census Bureau, employs thousands of people and spends millions of dollars each year just to collect and analyze data. In many ways statistics is being used to learn what is happening today and what tomorrow may be like.

EXERCISE SET 1
Uses of Statistics

1. Find some news items or advertisements which use data to describe our world of today:

a. in sports	e. in politics
b. in weather conditions	f. in science
c. in crop reports	g. in school
d. in business conditions	h. in government agencies

2. What are some problems in your community that need a collection of data to determine a course of action?

Using Data to Find Answers

Township High School was a new school with 200 students enrolled in its classes. Shortly after the opening of the school, a survey was conducted to determine how the students came to school. The following results were obtained:

Walked	90 students
Drove their own cars	20 "
Rode in other persons' cars	18 "
Needed bus service	72 "

On the basis of these data, the school made adequate provisions for student parking, contracted for three special buses, and set up a system of traffic control.

This is an example of the kind of information that your school board, superintendent and school architect needed in planning your school. They had to answer questions like these. Is garage space needed for school buses? How much parking space is needed? How many students go home for lunch? How much time is needed to travel home? How much traffic control is needed at the corner? Should students be permitted to drive to school? Accurate data is needed to answer these questions.

In the above example, the set of data that answered a question was used as a basis for some important decisions. In this case, it was easy to have the data include all the students in the school, and it was easy to use this set of data in the form in which it was collected. But if you wanted to carry out a similar investigation in a school with as many as 2,000 students, you might want to find some short cuts for collecting, analyzing, and presenting the data. If you were doing this investigation for a large city having many schools, you would probably want to survey only part of the students in the school system. In other words, you would want to work with a *sample* of the students from the schools. In statistics, a sample is a group of objects, scores, or persons selected from a larger group that is under consideration.

If you were going to use samples in a school transportation survey, you might get different results depending upon the location of the school, the enrollment of the school, the weather conditions on the day the samples were taken, and the way the students interpreted the questions asked in the survey. Thus, you would have the problem of selecting the proper samples.

You can see that the information you get from samples might be uncertain. Decisions made must take into account how, when, and where the sampling was done, how the sample was selected, and how large the sample was. Even then, there is uncertainty about the actual situation. Every decision made on the basis of samples has some uncertainty attached to it. Although we usually think of mathematics as being very certain in its conclusions, we will show later how mathematics gives us ways of making decisions based on uncertain data.

By examining the example given in this section, let's see what steps are needed to solve a statistical problem.

First, we find a problem that needs a solution — in this case, "How do students travel to and from school every day?" Next, we decide how to collect information about the problem. Do we need a sample? If so, how large should the sample be and how should it be selected? Then, we collect the necessary data, record it, and summarize it. In order for it to be readable, we might present the data in a table or a graph. We then analyze and interpret the data. Finally, we make a decision for the entire group on the basis of the sample results. When we do this, we must recognize the uncertainties, the risks, and the probability of being right or wrong about the entire group of students.

You can see that using statistics to solve problems can involve many things. In some cases, it is possible to use exact data to solve problems. In other instances, samples must be used, giving uncertainty to the results. In almost all cases it is desirable to have some methods to help organize, analyze, present, and interpret the data. Thus, if we are to understand and use statistics, we will have to be able to work with such things as tables, averages, graphs, sampling, and the laws of chance.

EXERCISE SET 2
Collecting Data

What data should be collected to get answers to these problems? In which problems would you use samples? Suggest some ideas about how you would select your samples.

1. What is the gas mileage for new Fords?

2. How many left-handed desks should be ordered for every classroom in the new school in your town?

3. Why are students tardy for third-period classes?

4. How much time do ninth graders spend watching TV during a week?

"Figures Don't Lie, But Liars Use Figures"

Just as there are many ways of using statistics to find new facts, there are also many ways of misusing statistics. Sometimes statistics are actually used to tell lies. Advertisements sometimes use statistics to make a product sound attractive or effective. For example, a toothpaste advertisement says, "Seven out of eight dentists use this toothpaste." But the ad never tells you how these eight dentists were chosen or who they are. A patent-medicine advertisement says that experiments prove that the medicine kills millions of bacteria in a few minutes. But what kind of bacteria does the medicine kill? Maybe they are harmless bacteria or even helpful bacteria rather than the kind that cause disease. We might also ask about the conditions under which the medicine kills these bacteria. Even sunshine will kill millions of bacteria in a few minutes. So maybe sunshine is a more effective germ killer than this patent medicine!

Suppose an ad says that seven out of ten farmers report that a certain cheese is twice as tasty as another cheese. How is taste measured so that you can report that something tastes twice as good? How, when, and where was the sample of ten farmers found?

Some people say that statistics show that teen-agers get into more trouble now than they did in "the good old days." These people usually quote a report about crime or auto accidents as proof. Maybe teen-agers are bad these days, but before you decide that they are worse now than they were in your parents' generation, you must compare the data of the two generations. Perhaps there are more bad teen-agers today, only because there are more teen-agers.

If you know something about statistics, you will not be fooled by false statements or clever ads based on wrong interpretations of data. You will understand that the praise of a product by a small, selected sample does not *prove* that the product is superior. You will know that the opinion of the "average" citizen is not always the "right" opinion. And you will know that data for one generation of people must be compared with comparable data of other generations before conclusions can be made. Whenever you see a statement based on statistics, you should ask questions about who collected the data, the source of the data, the method of collection of the data, and the real meaning of the data.

EXERCISE SET 3
Analyzing Data

What is wrong with the conclusions based on the data given in problems 1 through 5?

1. How many days do you go to school?

	Days
Days in a year:	365
You sleep at least 8 hours per day or $\frac{1}{3}$ of the year:	− 122
This leaves:	243
You have 52 Saturdays and 52 Sundays off:	− 104
This leaves:	139
You have summer vacation for three months:	− 90
This leaves:	49
You have Christmas and Easter vacations:	− 19
This leaves:	30
And you spend at least 2 hours each day eating:	− 30
Days left to go to school:	0

2. More people were killed in airplane accidents in 1960 than in 1928. Therefore, it was more dangerous to ride an airplane in 1960 than in 1928.

3. Checkered cows produce 26 per cent more milk than other cows. Therefore, checkered cows are the best milkers.

4. There are fewer accidents in France than in Germany. Therefore, it is safer to drive a car in France than in Germany.

5. Everybody who used Septa got over his cold in seven days. Therefore, Septa is a cure for colds.

6. Suppose that you are offered a job with two plans for salary increases:

    a. $3,000 per year, with a $200 increase every six months.

    b. $3,000 per year, with a $500 increase every year.

Which salary plan would you pick? How did you make your decision?

7. Find several news items, articles, or editorials which are based on current data. Have the data been used clearly and correctly?

8. Find several ads which refer to experiments or statistics. Are the statements true, false, or questionable?

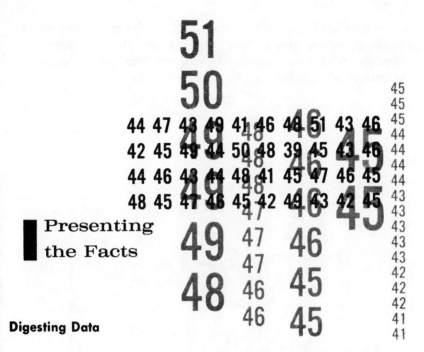

Presenting the Facts

Digesting Data

We have seen that it is often possible to use data to solve a problem, but we have also seen that data can be misinterpreted easily. Apparently we need some methods and ideas that will help us organize and interpret data.

Suppose you had a score of 45 on a mathematics test. Is this a high score or a low score? You don't know what this score means until you know the scores of others who took this same test. A score or measurement by itself means very little. We need to know how it compares with other scores.

Suppose that these are the scores in the mathematics test:

44, 47, 43, 49, 41, 46, 48, 51, 43, 46,
42, 45, 49, 44, 50, 48, 39, 45, 43, 46,
44, 46, 43, 44, 48, 41, 45, 47, 46, 45,
48, 45, 47, 46, 45, 42, 49, 43, 42, 45.

Even a large set of numbers like these 40 scores do not tell you much when presented in the above way. As you examine these 40 scores, you may be able to find the highest score or the lowest score; but to get a better understanding of the scores, you have to rearrange them in some way. One way to do this is to rewrite the scores in the order of their size, from the highest to the lowest, and then show how often each score occurred. The tally marks in Table 1 below show how this is done. The number of times each score appears is called the *frequency* of the score. This kind of summary of data is called a *frequency distribution*.

Table 1

Scores on Math Test

Score	Tally marks	Frequency (number)
51	/	1
50	/	1
49	/ / /	3
48	/ / / /	4
47	/ / /	3
46	//// /	6
45	//// / /	7
44	/ / / /	4
43	////	5
42	/ / /	3
41	/ /	2
40		0
39	/	1

The total number of scores is 40. This total is often represented by the symbol N. You can now get a better idea of where you rank in class with a score of 45.

Sometimes the data has scores that differ so greatly that it is necessary to collect similar scores into groups. Table 2 lists the scores on a physical education class weight-lifting test (given in pounds lifted).

Table 2

Scores on a Weight-lifting Test (in pounds)

129.6	60.3	80.2	93.0	100.0	136.4	92.1	122.6
91.8	102.0	129.4	99.1	103.6	78.0	91.0	139.1
92.0	87.6	72.9	153.4	75.5	114.0	82.3	98.0
93.0	61.9	86.0	81.1	96.0	75.6	84.2	108.3
48.0	107.7	72.3	81.4	100.2	28.5	103.6	95.2
83.3	62.1	102.1	96.0	77.7	96.6	57.6	118.2
56.6	38.1	63.3	64.4	81.0	66.6	88.1	
58.0	110.1	43.3	115.0	65.1	116.2	59.4	

Since the distribution of these scores would require many classifications, it is helpful to group the scores. The size of the group is called the *interval* for the group. In this example it seems sensible to use an interval of ten pounds. Table 3 is a frequency distribution of the scores from Table 2 using an interval of ten units (pounds).

Table 3

Scores on a Weight-lifting Test (in pounds)

Weight lifted interval	Group boundaries	Midpoint of interval	Frequency
150 — 159	149.5 — 159.5	154.5	1
140 — 149	139.5 — 149.5	144.5	0
130 — 139	129.5 — 139.5	134.5	3
120 — 129	119.5 — 129.5	124.5	2
110 — 119	109.5 — 119.5	114.5	5
100 — 109	99.5 — 109.5	104.5	8
90 — 99	89.5 — 99.5	94.5	12
80 — 89	79.5 — 89.5	84.5	10
70 — 79	69.5 — 79.5	74.5	6
60 — 69	59.5 — 69.5	64.5	7
50 — 59	49.5 — 59.5	54.5	4
40 — 49	39.5 — 49.5	44.5	2
30 — 39	29.5 — 39.5	34.5	1
20 — 29	19.5 — 29.5	24.5	1

Scores are tabulated in the intervals by following the rules of measurement. For example, a weight of 129.6 is closer to 130 than to 129 and thus would be tabulated in the interval 130-139. Similarly, a weight of 129.4 would be tabulated in the interval 120-129. The actual boundaries of the interval 120-129 are 119.50-129.49. This is usually written 119.5-129.5. We would put a boundary score such as 129.5 in the higher group, namely 129.5-139.5.

Sometimes the midpoint of the group interval is used as the representative of the group. Since the 150-159 interval includes scores from 149.5-159.5, the midpoint of this interval would be a score midway between 149.5 and 159.5, or 154.5.

Whenever data are tabulated in intervals, some information is lost. For example, in Table 3 there are two scores in the interval

120-129. We do not know the exact weights of these two scores just by looking at Table 3. We know only that they lie between 119.50 and 129.49 pounds.

EXERCISE SET 4
Working with Frequency Distributions

1. Below are scores on a mathematics test. Arrange them in order from highest to lowest. What is the middle score? What is the third from the top? What per cent of the scores are below 32?

$$19, 21, 36, 25, 32, 23, 28, 20, 34, 28, 31, 33$$

2. Tally the following scores in a frequency distribution. Do not use grouping.

$$92, 96, 87, 93, 92, 90, 97, 89, 86, 90, 91, 95,$$
$$88, 90, 87, 91, 93, 90, 93, 92, 95, 91, 88, 90$$

3. The following frequency distribution gives the heights of a group of tenth-grade students.

Height interval (in inches)	Frequency tally
73 — 74	/
71 — 72	/ /
69 — 70	++++ /
67 — 68	++++
65 — 66	++++ / / /
63 — 64	++++ ++++ / /
61 — 62	/ / / /
59 — 60	/ /
57 — 58	
55 — 56	/

a. What is the most common height of these tenth graders?
b. What is the tallest measurement possible to be recorded in this table?
c. In what interval would the measurement 62.6 inches be recorded?
d. How many students are shorter than 64.5 inches?

4. Copy this table and fill in the blanks.

Recorded interval	Real score boundaries	Units in interval	Midpoint of interval
a. 73 — 77	72.5 — 77.5	5	75.0
b. 70 — 74	_____	___	_____
c. 70 —	_____	10	_____
d._____	_____	3	75.0

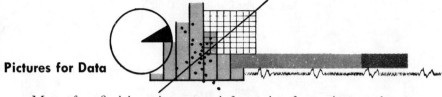

Pictures for Data

Most of us find it easier to get information from pictures than from written material. It is possible to present a great amount of information in a photograph or a drawing. Mathematicians make use of these facts by picturing numerical relationships with graphs. Graphs are very useful in the field of statistics, for they can picture numerical facts and clearly and quickly summarize data relationships.

There are several kinds of graphs we can use for picturing data. These include *circle graphs*, *rectangular distribution graphs*, *bar graphs*, *dot frequency graphs*, *histograms*, *line graphs*, and *frequency polygons*. In presenting data with a graph, our first task is to select the proper type of graph to represent the data at hand. Then we must give the graph a title and labels so that it can be read and interpreted with ease.

Consider the data in Table 4. The third column shows that 94% of the total number of drivers in fatal accidents are men and 6% are women. This indicates that, for every 100 drivers involved in fatal accidents, 94 are men. This expresses a comparison of men drivers to the *total* number of drivers. Comparisons of this type can be pictured with *rectangular distribution graphs* or *circle graphs*, as shown in Figure 1. The area of each region of each graph is drawn proportional to the data. In Figure 1, the portion of each graph showing women drivers is 6% of the total area.

In the circle graph, the per cent is converted to degrees so that the circle can be properly divided; for example, 6% of 360° = .06 × 360° = 21.6°, rounded off to 22°. Thus, a 22° arc portion of the circle is used to represent the portion of women drivers involved in fatal accidents.

Table 4
Drivers in Accidents

	Number of drivers in fatal accidents	Per cent
Male	36,700	94
Female	2,300	6
Total	39,000	100

Rectangular Distribution Graph

Drivers in Fatal Accidents

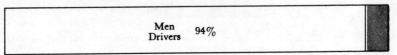

Women
Drivers
6%

Circle Graph

Drivers in Fatal Accidents

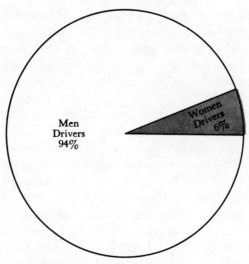

Figure 1

We must be careful in interpreting these graphs. For example, they do not indicate that women drivers are better than men drivers. For such a conclusion, we would need further information, such as the number of miles driven both by men and by women, the time and place of the driving, and the total number of men drivers and the total number of women drivers.

The expected scores on a standard mathematics test for students in grades 7, 8, 9, and 10 are listed in Table 5. Data of this type can be easily illustrated with a *bar graph*, as shown in Figure 2.

Table 5

Grade	Expected test scores
7	9
8	12
9	20
10	25

Expected Scores on Standard Mathematics Test

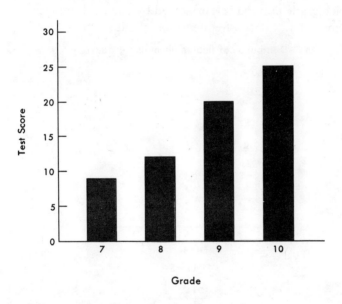

Grade

Figure 2

In a bar graph, the height of each bar is drawn in proportion to the number or size of the measures, scores, or percentages. The bars are of uniform width and the spaces between bars should be the same as the width of the bars. The vertical and horizontal scales (sometimes called axes) should be labeled completely. On the horizontal scale it is usual to enter scale values at the middle of the bar.

A frequency distribution in which the data has not been grouped in intervals can be represented with a dot frequency graph. The distribution of final grades in a mathematics class is given in Table 6.

Table 6

Grade	Frequency
A	3
B	5
C	9
D	2
F	2

The data of Table 6 is pictured in Figure 3. The number of times each grade occured is represented by dots — 1 dot represents a frequency of 1, 2 dots a frequency of 2, and so on.

Distribution of final mathematics grades

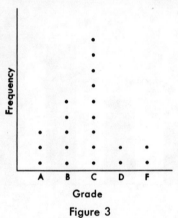

Grade

Figure 3

A frequency distribution for the total number of runs scored by thirty-eight Little League teams during an eight-game schedule is given in Table 7. This distribution is pictured with a *histogram* in Figure 4. The histogram is similar to the bar graph, but there are no spaces between the bars. It is usually used to picture frequency distributions. In a histogram, the group boundaries are marked off with a uniform scale on the horizontal axis, and the frequencies are shown on the vertical axis. The horizontal scale in Figure 4 is marked with the group boundaries 17.5, 19.5, 21.5, and so on. With this type of marking, the histogram definitely shows that scores of 24 and 25 are in the group having a frequency of 6, that scores of 26 and 27 are in the group having a frequency of 12, and so on. In Figure 4, a break is shown on the horizontal axis to indicate that the scale is not uniform between 0 and 17.5.

Table 7

Score interval	Group boundaries	Frequency
32 — 33	31.5 — 33.5	2
30 — 31	29.5 — 31.5	5
28 — 29	27.5 — 29.5	8
26 — 27	25.5 — 27.5	12
24 — 25	23.5 — 25.5	6
22 — 23	21.5 — 23.5	4
20 — 21	19.5 — 21.5	1

Total Runs Scored During Little League Season

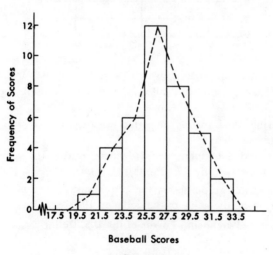

Figure 4

A line graph, sometimes called a *frequency polygon*, is also used in statistics to graph a frequency distribution. It is like the histogram in its vertical and horizontal scales. It is drawn by connecting the midpoints of the upper bases of the histogram bars, as has been done with the broken line in the graph in Figure 4.

There are many types of statistical line graphs that are variations of the frequency polygon. Problems 1 and 2 of Exercise Set 5 are examples of some other types of line graphs.

1. The following graph shows the miles traveled by automobiles in the United States for various years from 1936 to 1954.

Automobile Travel in the U.S.

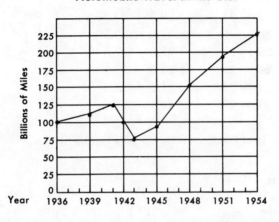

a. How many more miles were driven in 1954 than in 1942?

b. If 36,000 persons were killed in 1936, what were the miles driven per death?

c. During which three-year period was there the greatest *increase* in miles traveled?

d. Why did automobile travel decrease during the years 1941 to 1943 as shown by the graph?

2. The following graph shows the purchasing power of the United States dollar, using the value of the dollar in 1945 as a base.

Purchasing Power of the U.S. Dollar
1945-1956

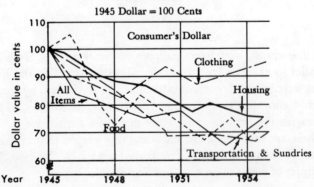

a. What commodity has increased most in price? Least?

b. When did prices on all items increase most rapidly? What is the probable reason for this increase?

c. One dollar spent on clothing in 1951 would buy the same as what amount spent on clothing in 1945?

d. Clothing worth $.90 in 1945 would cost how much in 1951?

e. Clothing worth $1.00 in 1945 would cost how much in 1951?

f. About how much would it cost to build a house in 1954 that would have cost $10,000 in 1945?

3. Thirty-five test scores are given in the table below. Complete the frequency distribution columns and then construct a combined histogram and frequency polygon for the distribution.

21	26	21	20	23	24	22
19	24	26	25	23	26	29
21	24	19	25	26	25	22
25	27	23	26	24	25	30
25	23	27	24	28	28	28

Score interval	Group boundaries	Frequency
29 — 30	28.5 — 30.5	2

Graphs Don't Lie, But Liars Use Graphs

Although graphs can be very useful, sometimes they can be used to create false impressions. Here is a sample of two ways of picturing data.

Have average scores in football games increased a great deal? Figure 6 contains two graphs that use the same data to create different impressions.

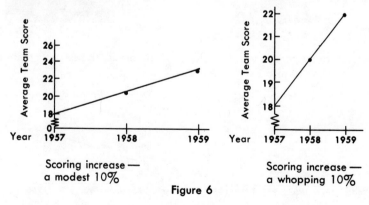

Scoring increase —
a modest 10%

Scoring increase —
a whopping 10%

Figure 6

These two line graphs picture the same data. But the choice of scale for showing the changes gives different pictures and the titles exaggerate the difference in the scales.

When working with graphs, we must be very careful not to misinterpret them. The book *How to Lie With Statistics* by Darrell Huff contains many illustrations of how graphs may be misused to tell a lie. You should learn how to draw, read, and interpret graphs so you won't be fooled by graphs that are poorly made.

EXERCISE SET 6
Thinking About Graphs

1. Do the following graphs indicate that Country A spends more money on national defense than Country B? What other facts might be helpful?

Part of Budget Devoted to National Defense

2. Do the following histograms indicate that Class I is more intelligent than Class II? Why or why not?

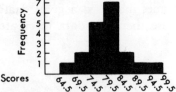

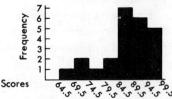

3. According to the graphs below, which brand shows the greatest yearly increase in price?

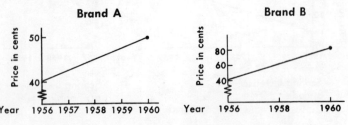

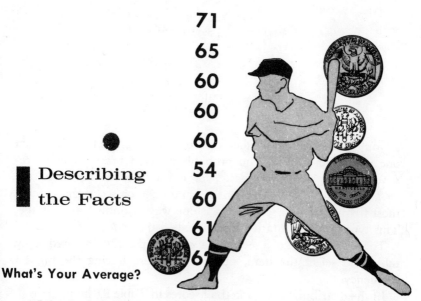

71
65
60
60
60
54
60
61
6
6

Describing the Facts

What's Your Average?

"John had three hits out of four times at bat. He certainly was hitting above his average."

"I certainly had a below-average score on the last math test."

"The average weekly allowance was seventy-five cents."

You have undoubtedly heard comments similar to the ones listed above. Note that all the comments contain the word "average," but in the three situations given, the word probably has different meanings. When you see the words "average score," you probably think of other terms, such as "middle score," "most likely score," "score that occurs most often," and so on. Averages are sometimes called "measures of central tendency," because they tell us the score or number in a situation that we expect to occur very often. Such a score will "tend" to describe all the data. Just as the word "average" has many interpretations, there are several mathematical ways of finding representative, typical, or average scores. In statistics, the word "average" is used to refer to three different measures of central tendency.

Mode or "Most"

Suppose you own a drugstore and sold different brands of toothpaste for the past month as shown in Table 8.

Table 8

Brand	Number of tubes sold
A	6
B	7
C	18
D	34
E	10

If you could continue to stock only one of these five brands, which one would you select? You probably selected Brand D as representative of a good toothpaste, for it was the one that was sold *most* frequently. The selection of a number or score that occurs most frequently in a set of data is one way of finding an average. This type of representative score is called the *mode*.

To find the mode of a set of data, arrange the data in a frequency table and select the item, number, or score having the highest frequency.

In the distribution of basketball scores in Table 9, the mode is 8. More players (5) had a score of 8 than any other score.

Table 9

Score	Frequency
10	1
9	3
8	5
7	4
6	2

Median or "Middle"

The hits made by the players on a junior-high-school baseball team during a season are shown in Table 10. What number would represent the number of hits for an "average" player on the team?

Table 10

Bob	10	Mike	34	Fred	5
Jim	9	Joe	3	Paul	8
Bill	4	Tom	5	Art	11
Harry	4	Walter	7		

The mode would probably not be a good average in this case, for there is no one value that would meet the definition of mode.

However, we could get an idea of the hits for an average player on the team by arranging the numbers of Table 10 in order,

$$34, 11, 10, 9, 8, 7, 5, 5, 4, 4, 3,$$

and then selecting the middle number. Since there are eleven numbers in the list, the middle number is 7, for there are five numbers on either side of it in the list. Such a measure of central tendency, that selects the middle number as being representative of a set of numbers, is called the *median*.

To find the median of a distribution, first arrange the scores in order, and then find half the number of scores to determine the position of the middle score. If there are an odd number of scores, add 1 before you find half the number. Count up from the lowest score or down from the top score until you get to the middle score. If the distribution has an even number of scores, the median is found halfway between the two scores closest to the middle. The median of the five scores 13, 14, 15, 18, 21 is 15. The median of the six scores 42, 45, 50, 54, 57, 58 is halfway between 50 and 54, or 52. The median is easy to compute and is not influenced by a few extremely high or low scores.

Mean or Arithmetic Average

The students in a high-school science class were testing a new type of flashlight battery. They tested for the operating life of nine batteries and obtained the following results: 20, 21, 22, 22, 22, 28, 29, 29, 32 hours. The class wanted to give a statement of the average life of the batteries tested. They did not want to use the mode or median as an average for the data. They felt it would be much better to use a value that, when used nine times in a sum, would give the same result as the sum of the nine values obtained in their experiment. To find such a value, they added the values from the experiment and divided by 9, the number of values, obtaining 25 as a result. The students felt that this was a better representation of the average life of the batteries than the 22 obtained from the mode or median, for this value took into account all the individual values in the experimental data. This measure of central tendency is known as a simple arithmetic average and is usually called the *mean*. The mean is often represented by the letter M.

As we have already seen, to find the mean of a set of scores we add the scores and divide by the number of scores. For example, the mean for the scores 14, 17, 18, 19, 22 is

$$M = \frac{14 + 17 + 18 + 19 + 22}{5} = \frac{90}{5} = 18.$$

The formula that tells you how to compute the mean is

$$M = \frac{\Sigma X}{N}.$$

Here the symbol Σ is the capital Greek letter S and is read "sigma." The symbol Σ is a "summation sign" and always indicates addition. Usually X is used to represent single scores such as 14, 17, 18, and so on. Thus ΣX says, "Find the sum of all the scores." N represents the number of values in the data. Therefore, $M = \frac{\Sigma X}{N}$ says, "The mean of a set of data is equal to the sum of all the data values divided by the number of values."

EXERCISE SET 7
Some Easy Averages

1. Find the mode of each of the following sets of data:
   a. 2, 3, 4, 7, 5, 7, 3, 3, 10, 4, 9
   b. 24, 21, 20, 25, 21, 27
   c. 3, 3, 1, 2, 2, 3, 2, 4, 5, 4, 3, 4, 2, 3, 3, 4
2. Find the median of each of the sets of data given in problem 1.
3. Find the mean for each of the sets of data given in problem 1.

Averages from Frequency Distributions

We have seen that frequency distributions are very useful in organizing large sets of data. Therefore, it is important to be able to compute averages from a frequency distribution.

In a frequency distribution in which the data is *not grouped* in intervals, the mode and median are found in the usual way. In a frequency distribution in which the data *is grouped* in intervals, we use some new methods to compute the mode and the median.

The mode in such a frequency distribution is the *midpoint* of the interval with the greatest frequency.

In computing the median, we find the interval in which the middle score occurs. Then we mathematically determine a value from the interval that should correspond to the middle score.

Consider the example having the data given in Table 11.

The mode is 62, the midpoint of the interval with the greatest frequency.

Table 11

Interval	Midpoint	Frequency
75 — 79	77	1
70 — 74	72	3
65 — 69	67	5
60 — 64	62	10
55 — 59	57	4
50 — 54	52	2
		$N = 25$

The middle score is the thirteenth. This score is found in the interval 60-64. Counting from the bottom of the table we find 6 values below this interval. We need seven more values to reach the thirteenth score. This means we need 7 out of the 10 values in this interval to get to the thirteenth score. The length of this interval is from 59.5 to 64.5, or 5 units. If we assume that the scores are evenly distributed within the interval, the score that corresponds to the median would be $\frac{7}{10}$ of the way between 59.5 and 64.5. We need $\frac{7}{10}$ of this interval; $(\frac{7}{10}$ of 5$) = 3.5$. Adding 3.5 to the lower boundary of this interval $(59.5 + 3.5)$ gives 63 as the median.

A short way to find the mean when scores are grouped without intervals is to multiply each score by its frequency, add these products, and then divide by the number of scores. The formula representing this method is written $M = \frac{\Sigma f X}{N}$, where f refers to frequencies.

This method is used in the work in Table 12.

Table 12

Scores: 8, 7, 8, 9, 6, 10, 8, 9, 6, 9, 8, 7, 10, 8, 7

Score (X)	Frequency (f)	fX
10	2	20
9	3	27
8	5	40
7	3	21
6	2	12
$\Sigma f = N = 15$		$\Sigma fX = 120$

$$M = \frac{\Sigma fX}{N} = \frac{120}{15} = 8$$

When finding the mean for a frequency distribution grouped in intervals, the scores are considered to have the value of the midpoint of the interval. Table 13 illustrates a problem of this nature.

Table 13

Interval	Midpoint (X)	Frequency (f)	fX
42 — 44	43	2	86
39 — 41	40	0	0
36 — 38	37	3	111
33 — 35	34	5	170
30 — 32	31	8	248
27 — 29	28	6	168
24 — 26	25	4	100
21 — 23	22	1	22
18 — 20	19	1	19
		$\Sigma f = N = 30$	$\Sigma fX = 924$

$$M = \frac{\Sigma fX}{N} = \frac{924}{30} = 30.8$$

When using the above methods for computing averages for grouped data, we cannot expect the results to be exactly the same as those that would be obtained if the scores were considered individually. We assumed either that all the scores were concentrated at the midpoint of the grouping interval or that the scores were evenly distributed throughout the interval. Of course, these conditions would be highly unlikely. However, the results obtained from methods for computing averages with grouped data are accurate enough to justify the use of the methods.

EXERCISE SET 8
Finding Averages

1. Find the mode (most frequent score) of each of these distributions.

a. 86, 82, 78, 93, 86, 84, 81, 90, 85, 79, 86, 85, 88, 81, 87

b.

Score	Frequency
24	2
23	3
22	5
21	8
20	4
19	3
18	1

c.

Interval	Midpoint	Frequency
60 — 64	62	1
55 — 59	57	3
50 — 54	52	2
45 — 49	47	5
40 — 44	42	8
35 — 39	37	10
30 — 34	32	7
25 — 29	27	4
20 — 24	22	3
15 — 19	17	0
10 — 14	12	1

2. Find the middle score (median) of each of the distributions in problem 1 above.

3. Find the mean of each of the distributions below by the long method of adding all scores.

a. 9, 12, 7, 6, 8, 11, 3

$$M = \frac{\Sigma X}{N} \qquad N = ? \qquad \Sigma X = ? \qquad M = ?$$

b. $- 5, 3, - 1, 0, 4, - 7, - 4, 2, - 3, 6$

c. Add a score of 32 to the distribution in *a*. Now what is the mean? This illustrates how one extremely high score can raise the mean.

4. Find the mean of each of the distributions of problem 1 above by multiplying each score by its frequency (f) and then summing (ΣfX).

a. $\Sigma fX = ?$ b. $\Sigma fX = ?$ c. $\Sigma fX = ?$

     $N = ?$ $N = ?$ $N = ?$

     $M = ?$ $M = ?$ $M = ?$

Comparing the Three Averages

What's wrong with the following statement? "The average wage at our factory is $5,800." The difficulty here is that we don't know what measure of central tendency this is. The statement was made by the factory owner, whose salary is $47,000 per year. The labor union representing the factory workers said the average wage was $4,000. The tax agent representing the Internal Revenue Bureau of the Federal government said the average wage was $4,766. These different answers were all obtained from the data in Table 14.

Table 14

Annual wage	Number receiving this annual wage
$47,000	1
17,000	1
12,000	2
7,000	3
5,000	11
4,000	30
3,000	2

The three measures of central tendency are:

Mean wage   = $5,800
Median wage = $4,766
Mode wage   = $4,000

These measures can be interpreted as follows: the mean or arithmetic average indicates that if the money was distributed so that each person was paid the same, each employee would get $5,800. The mode tells us that the most common annual wage is $4,000. The median indicates that about half the employees get less than $4,766 and half get more than that amount. The example above shows how the mean, median, and mode give different numbers that mean different things. So when you see an average reported, ask, "What average?" "Who is included?" and "How accurate are the data?"

The major purpose of an average is to state a single score which is typical of all the scores in the set of data.

The arithmetic average, or mean, is usually considered the best measure of central tendency, and it is undoubtedly the most common. It is easily expressed by a formula, and it takes into account every score in its computation.

The arithmetic average or mean may not be typical if the data include a few extreme values in one direction.

Example: A classroom group of five people in a high school averages 25 years in age. Four are students, one a teacher. Is 25 a typical age in the group? What is the average if students are all 16 years of age, the teacher 61? Here we note that one high score has too great an influence on the results.

The mean is used in many situations such as the following:
— in meteorology, for obtaining the average temperature or rainfall;
— in medicine, for discovering the average duration of a disease;
— in anthropology, for estimating certain average characteristics of a group of human beings;
— in business, for computing average wages, prices, index numbers, and so on.

The mode is often thought to be the most typical score of all (since it is the most frequently occurring), but it does not take account of the other values in the data. It is easy to determine, but there are often several values from a set of data that meet the definition of the mode. However, sometimes the mode alone is the appropriate average.

Example: A sport shirt manufacturer has equipment to make only one size of a man's sport shirt, and he must choose the size to make. He decides on the arithmetic-mean size of shirt bought by men. To sell more shirts, he should have chosen the mode (the most common size) to manufacture.

The median is the middle score and is not influenced by the other values in the data except as they are either above or below the median. That is, so long as a score is below the median, it doesn't matter whether it is just barely below or an extremely low score. However, if scores are concentrated in distinct and widely separated groups, the median may be of little value as a measure of central tendency.

Example: On a 30-point test, there are five scores of 20, three scores of 26, and one score of 29. Using the median, we would have to say that an average score in the class was 20. However, we question this as being representative of the class because of the peculiar way the scores are distributed. In this case, the mean would probably provide a better measure of central tendency.

The median is used in many statistical investigations, including the following:
— in insurance, for finding the average length of life;
— in the study of drugs, for finding the potency of a drug;
— in industry, for testing the quality of a product.

EXERCISE SET 9
Selecting the Proper Average

1. The following prices for 6-ounce jars of silver polish are obtained from nine different manufacturers: 20¢, 20¢, 22¢, 24¢, 24¢, 36¢, 38¢, 42¢, 45¢. What value would you give as an average price for silver polish? Which average did you select? Why?

2. The following contributions for the scholarship fund of a small college were obtained after contacting thirty alumni of the college:

Amount (in dollars)	Frequency
1,000	2
100	2
80	2
50	3
30	4
10	6
5	8
0	3

a. Compute the three averages for this data.
b. What is the "usual" contribution?
c. Which average would best represent these data? Why?

3. Suppose you are the buyer (the person who selects the articles to be sold) for a department store. Which type of average will probably be of the greatest value to you? Why?

4. Twenty pupils are given a test that is to be used to determine the placement of pupils in good and fair sections of a senior mathematics course.

a. If the scores are 66, 67, 67, 69, 70, 70, 72, 73, 74, 76, 85, 86, 88, 88, 90, 92, 94, 97, 98, 99, which average would be the best one to use to help determine the placement of the pupils?

b. If the scores are 62, 63, 66, 66, 66, 66, 67, 68, 68, 69, 70, 87, 89, 90, 92, 95, 98, 98, 99, 100, which average would be the best to use to determine the placement of the pupils?

318

5. A tire manufacturer tests twenty-four of his tires to determine the number of miles they can be driven before they wear out, and obtains the following results:

Miles	Number of tires
60,000	1
55,000	1
35,000	6
30,000	8
25,000	8

Which average would be the most appropriate one for describing the average life of the tires produced by this manufacturer? Why?

How Do You Rank?

Robert is five feet tall and had a score of 56 in his mathematics test. Last Saturday he earned $1.50. Is Robert a short boy? Is he good in mathematics? Is he poorly paid? None of these questions can be answered from the data reported. These numbers have meaning only as they are compared with averages or standards. The standard for comparison is usually furnished by the group to which the person belongs. If Robert is nine years old, he would be considered a tall boy. If the median score of nine-year-olds on the mathematics test is 60, he is not superior on this test. If he earned the $1.50 by working half an hour in the garden, his rate of pay would be very high.

A measurement usually has meaning only when we compare it to a group of similar measurements. One way to show comparison is to give the rank of the score from the top. A typical use of rank is to say that Bill ranks seventh in his class. However, rank has little meaning unless we know how many are in the total group.

If Bill is in a class of seven students, he ranks at the bottom. If he ranks seventh in a class of 350, he has a high rank. Thus, a better way to give Bill's relative standing is to get some type of comparison between Bill's rank and the total number of people in the class. Let's use the division operation to compare the number of people below Bill to the total number of people in the class.

If Bill is ranked seventh in a class of 350 students, there are $350 - 7$ or 343 students ranked below Bill. If we compare 343 to 350 by division, we have $\frac{343}{350}$ or $\frac{49}{50}$. We can change this comparison to a per cent by writing it as $\frac{98}{100}$ or 98%. We say that Bill's *percentile rank* (*PR*) in his class is 98, which means that ninety-eight people out of every one hundred in Bill's class are ranked *below* him. A percentile rank tells us the per cent of persons in a group that are below a certain score, position, or rating.

Suppose Robert's score of 56 in the mathematics test is one of seventy-two scores. If there are eighteen pupils with scores below 56, Robert's percentile rank is $\frac{18}{72} (100) = 25$. We could also say that a score of 56 is the 25th percentile, meaning that 25 per cent of the scores are below 56.

EXERCISE SET 10
Per cents and Percentiles

1. The scores of twenty students on a spelling test were the following: 2, 5, 6, 7, 8, 9, 10, 10, 11, 11, 11, 12, 12, 12, 13, 13, 14, 16, 17, 18.
    a. What is the rank from the top of a score of 14?
    b. What per cent of the scores are below 14?
    c. What is the percentile rank of a score of 14?
    d. What score has a percentile rank of 25 — in other words, what is the 25th percentile?
    e. What is the 40th percentile?

2. In a class of twenty-four pupils, Jane has a rank of 5. What is Jane's percentile rank in the class?

3. In this same class of twenty-four pupils, Charles has a percentile rank of 25.
    a. How many pupils had scores lower than Charles?
    b. What is Charles' rank in the class?

4. The following represents the distribution of height in inches of a ninth-grade class.

Height in inches	Frequency
66	1
65	2
64	1
63	3
62	5
61	8
60	7
59	9
58	4
57	5
56	1
55	3
54	1

Total 50

a. What is the rank of a height of 64 inches?

b. What is the percentile rank of a height of 64 inches?

c. What per cent were taller than 62 inches?

d. What is the percentile rank of Mark, who is 56 inches tall?

e. What per cent of the students are shorter than 62 inches but taller than 58 inches?

5. In a class of 100 seniors, Barbara has a percentile rank of 75. In this same class Gary ranks 75th. Who has the highest standing in the class? Copy and fill the blanks in this table:

	Rank	Percentile rank
Barbara	_____	75
Gary	75	_____

6. In a senior class of 100 students, Marjorie has rank 1 and Pinky has rank 100. Who has the highest standing in the class?

	Rank	Percentile rank
Marjorie	1	
Pinky	100	

7. John says he ranks 10th in his class at Washington High, while Judy says she ranks 20th at Central High. What additional information do you need before you can decide whether John exceeds Judy in school work?

Predicting Events

Sampling Is Tricky

Did you ever sample a turkey to see how it tasted? You couldn't eat the whole turkey or even a whole piece before dinner, so you used a sample. In the same way, Tom Sawyer cut a plug in a watermelon to sample its ripeness. These small samples of food usually predict the taste of the whole turkey or whole melon. In a similar fashion, governments, industry, schools, and businesses use samples to make statistical computations for the groups from which the samples were selected, predict events, and estimate the quality of products. Have you heard about ratings of TV shows, Hit Parade listings of favorite songs, and the Gallup Poll of public opinion? These are illustrations of the work of groups who obtain their information by sampling. If you stop to think, you will find that almost all the facts we know have been discovered by sampling. We know what mountains are like by seeing a few, not all of them. We know what poetry is like by reading some samples of poems. We learn to like symphony music by hearing samples of concerts or records.

Up to this point, most of our work with statistics has been concerned with representing, describing, organizing, and interpreting data that were very complete. In most of the examples we studied, we assumed that we had a complete list of the data under consideration. In many practical situations this is impossible. Hence, statisticians are often concerned with the selection of sample sets of data that can be used as representative of larger sets of data that cannot be completely obtained.

Whenever we get information from samples, we run the risk of making mistakes. The sample is not always like the whole thing. The sample of turkey may be only white meat, the poems you select may be poor ones, and the music you hear may be poorly played. But we are usually willing to take this risk of error by sampling. We don't want to eat the whole turkey to decide if it is delicious! We sample to save money, time, and the turkey.

In statistics, we learn how to pick out samples so that the risk of a mistake is small. Statistics also tells us how much risk we take of being wrong. We reduce the risk by using more than one sample, by using samples of large size, and by selecting the samples in a *random* fashion. A *random sample* is one that you pick by chance. For example, suppose you put the name of each person in your class on a card, put all the cards in a box, shake them until they are all mixed up, then reach in without looking and draw out five cards. Then you will probably have a random sample of five names from your class. In a card game, you usually shuffle the cards and deal them so that every player gets a random sample. Likewise, if you wanted to know the opinions of people in your town, you would draw a random sample from the names of all the people living in your town. You would not pick a sample of the people on the street, in the stores, or at church. Each of these places would give you a sample of a certain kind of people, not people in general, or at random. If you wanted to find out what make of car is most popular, you wouldn't ask only the people who sell Ford automobiles. The sample would include people having many different occupations and backgrounds so that you would get different opinions. Your sample should include people who do not own cars and people who own several different makes of cars if you want opinions that represent everybody.

Whenever you see a statement based on samples, you need to ask these questions:

How large was the sample?

Was the sample picked at random?

Can you apply the sample results to the entire group?

Be Cautious about the Samples

A common cause for the misuse of statistics is found in the samples that are to be used as the basis for statistical investigations.

There may be bias because of the method of selecting the sample.

Example: In the 1959 legislative session in Minnesota, the subject of daylight saving time was in controversy. A Minneapolis newspaper published a ballot for readers to indicate their preference as to daylight saving. Two days later, on the basis of ballots returned at that time, the newspaper announced that its readers favored daylight saving 2 to 1. The sample was probably biased because only those people who felt strongly on the matter were likely to return the ballot.

Erroneous conclusions may be reached if sample results are applied to a group different from the population sampled.

Example: Suppose a random sample in your high school showed that 76% of the students intend to go on to college. Would you be justified in concluding that 76% of all high school students in your state intend to go to college?

Sometimes inaccurate sample results are obtained because of the type of information requested, such as information which people do not easily remember or which they do not like to give.

Example: Would you get reliable information if you asked how many quarts of milk a family had bought during the last year? If you asked housewives how much money their husbands lost by gambling?

EXERCISE SET 11
Working with Samples

1. Find some advertisements that make statements based on sampling.

2. Give some examples of information about your city, state, or nation that is based on samples.

3. Why aren't these samples random?
   a. the members of the basketball team
   b. the people in a downtown store
   c. the people who live in your city
   d. the students at your state university

4. Why wouldn't the samples collected in the following ways include representatives of all opinions?

a. a questionnaire from all the members of the school band
b. a survey made by calling every tenth number in the yellow pages of a telephone directory
c. a survey by mail requesting a reply by mail
d. an interview of every tenth person met on a downtown street at noon.

What Are the Chances?

"Our basketball team has a 60-point scoring average per game. They should score about 60 points in the game Friday."

"I had a score of 86 on a standard mathematics test. This means that only 5% of the mathematics students in the country are ranked above me."

After checking 10,000 students from different schools in a large city, it was found that 400 students needed extra reading work. Since there were 50,000 students in all the schools of the city, the school board decided to set up special reading centers to handle approximately 2,000 students.

Statements such as the preceding ones are quite common. They all illustrate the use of statistical data in making decisions or predictions.

If the basketball team has averaged 60 points per game, does this mean that the team will score 60 points in the next game? Of course not. But we still might assume that there is a very good *chance* that this will be the case.

Does the test score of 86 mean that exactly 5% of the mathematics students in the country will rate above the student with that score? Certainly not. But past data indicate that such a condition will *probably* exist.

Does the reading check indicate that there will be exactly 2,000 students with reading difficulties? Of course not. However, if the sample of students was selected properly, it is *likely* that there will be about 2,000 students needing help.

Chance, probably, likely — all three words indicate uncertainties. In fact, we have seen many ways in which uncertainties can come up through the use of numerical information. We noted the risk involved when working with samples of data. Thus, *probability* or *chance* is a very important topic in statistics.

If we toss a coin 100 times, how many times will it come up heads? Of course, we really don't know, but we say that there is a better chance that it will come up heads fifty times out of the hundred chances than any other combination. We say that the probability for tossing a head is 50 out of 100, or 1 out of 2. We know that a coin will not *always* turn up heads 1 out of 2 times, but in the long run this is what we would expect.

This mathematical aspect of probability can be expressed in a more technical way. The probability of an event happening is defined as the ratio (a comparison of two quantities by division) of the number of ways an event can succeed to the total number of ways the event can occur. If p denotes the probability that an event will happen, f the number of favorable ways the event could happen, and t the total number of ways the event could happen, then a formula for numerically expressing the probability of an event happening is

$$p = \frac{f}{t}.$$

In tossing a coin, the coin can fall two ways ($t = 2$), but for normal coins, only one of these is heads ($f = 1$). Hence, the probability of getting a head when tossing a coin is numerically represented as $\frac{1}{2}$ or .5. If a two-headed coin is used, the probability of getting a head is $\frac{2}{2}$, or 1. A numerical probability of 1 indicates complete certainty. If a two-tailed coin is used, the chance of getting a head is $\frac{0}{2}$, or 0. A numerical probability of 0 indicates no chance of success. Therefore, we see that a probability is a number that must be greater than or equal to 0 and less than or equal to 1.

 What is the probability that a tossed die will turn up a 3? The number of ways a 3 can turn up on one die is 1. The total number of ways a die can fall is 6. Then the probability of turning up a 3 is $\frac{1}{6}$.

A set of four cards labeled 1, 2, 3, and 4 is placed in a box, and a second set of four cards labeled 1, 2, 3, and 4 is placed in a second box. If you are permitted to draw one card from each box (without looking, of course), what is the probability of drawing cards that will give you a total of 4? To answer this question, we have to know the number of different ways a total of 4 can be drawn and the total number of ways the cards can be drawn. Table 15 supplies this information. The possible drawings from the first box are indicated along the left side of the table, the possible drawings from the second box at the top of the table, and the possible combinations from both boxes in the body of the table.

Table 15
Second Box

	1	2	3	4
1	1,1	1,2	1,3	1,4
2	2,1	2,2	2,3	2,4
3	3,1	3,2	3,3	3,4
4	4,1	4,2	4,3	4,4

(First Box, along the left side)

The possible ways for drawing a total of 4 are (3,1), (2,2), and (1,3), or three ways. The total number of possible draws is 16. Therefore, the probability of drawing a total of 4 is $\frac{3}{16}$.

What is the probability of drawing a total of 3 *or* 4? The total number of possible draws is again 16, but since a favorable result will now be obtained from a total of 3 or 4, there are now five favorable combinations. Hence, the probability is $\frac{5}{16}$, or about .31. We could say that there is about a 31-out-of-100, or 31%, chance of drawing a 3 or a 4.

EXERCISE SET 12
Computing Chances

1. In the algebra class, there are eighteen boys and twelve girls. The names of all the class members are placed on cards in a box, mixed up thoroughly, and a card drawn without looking. What is the chance that the name drawn will be a girl?

2. In how many ways can two coins turn up? The possible ways are heads-heads, heads-tails, tails-heads, tails-tails.

  a. What is the chance that the two coins will turn up two heads?

  b. What is the chance that the two coins will turn up one head and one tail?

3. Show the different ways that three coins can turn up. There are eight ways in all.

  a. What is the chance that the three coins will turn up all heads?

  b. What is the chance that the three coins will turn up two heads and one tail?

4. A die is to be tossed once.

  a. How many different numbers could turn up?

  b. How many faces have only one dot on them?

  c. What is the ratio of the number of faces with just one dot on them to the number of ways the die could turn up?

  d. What is the probability of turning up a one if you toss the die just once?

  e. What is the probability of turning up a number other than a one if you toss the die just once?

5. Three white balls and four black balls are placed in a bag. All the balls are identical except for color. You are to make one drawing from the bag.

  a. What is the total number of balls in the bag?

  b. What is the ratio of the number of white balls to the total number of balls?

  c. What is the probability of drawing a black ball on one try?

6. What is the probability of drawing the seven of diamonds from a pack of fifty-two playing cards?

7. What is the probability of drawing any diamond from a pack of fifty-two cards?

8. What is the ratio of the number of aces to the total number of cards in a deck of fifty-two cards?

What, then, is the probability of drawing an ace in one draw?

9. What is the ratio of the number of face cards (kings, queens, and jacks) to the total number of cards in a deck of fifty-two cards?

What, then, is the probability of drawing a face card from a deck of fifty-two cards?

10. What is the probability of getting a total of 5 when throwing a pair of dice?

11. What is the probability of getting a 3 or a 4 when throwing two dice?

The Statistics of Scattering

Measures of central tendency such as the mean, median, and mode of a set of data give a single number which often presents a very incomplete picture of the data. This is illustrated by the following example. Examine the wage data in Table 16, and determine the mean wage in each plant.

Table 16

Annual wages	Employees in Factory A	Employees in Factory B	Employees in Factory C	Employees in Factory D
$10,000	1	5		3
9,000	1			1
8,000	1			1
7,000	1			
6,000	1		5	
5,000	1		5	
4,000	1			
3,000	1			1
2,000	1			1
1,000	1	5		3

For each of these distributions the mean wage is $5,500. However, the distributions differ greatly. A measure of the way the scores *scatter* is needed. Besides an average, we need a measure of the variation, scattering, spread, or dispersion of the data being considered.

A simple measure of dispersion is the difference between the largest and smallest scores of a set of data. This measure is called the *range* for a set of data. It is a poor measure of dispersion because it depends on only two scores, telling us nothing about the remaining scores. For example, the range in wages in factories A, B, and D is the same, $1,000 to $10,000, but you can see that the wages at these factories are scattered very differently. Nevertheless, the range does have many uses in the field of statistics, such as the quotation of temperature ranges in weather and climatic data.

Because the range has some shortcomings as a measure of dispersion, we obviously need a better method. Since the mean is the most common measure of central tendency, let's try to develop a measure of dispersion that is related to the mean.

Suppose the mean scores on two different sets of algebra tests (having the same number of possible points) are both 50. If you score a 70 on one test and a friend of yours scores a 74 on the other one, does your friend rank higher on his test than you do on yours?

Before you answer this question, it would be helpful to know how the scores on each test are scattered about their means. If most of the scores in your class are close to the mean, and if the scores in your friend's class differ greatly from the mean, your 70 is probably a better score than your friend's 74.

How could we get a measure of scattering about the mean of a set of data? Here is a very useful method. First, list each score (X) of a set of data. Then compute the mean (M) for the data. Next, find out how much each score varies from the mean by subtracting the mean from each score $(X - M)$. Since some of the deviations will be negative, square each deviation $(X - M)^2$ so that there will be no negative deviations to cancel positive ones. Now get the arithmetic average of the squares of the deviations by dividing the sum of the squares by the number of scores $\left(\dfrac{\Sigma(X - M)^2}{N} \right)$. Finally, undo the squaring by taking the square root of this average of the squares $\left(\sqrt{\dfrac{\Sigma(X - M)^2}{N}} \right)$. This represents a measure of the deviation of the scores from the mean.

This measure of dispersion is called the *standard deviation*, and tells how scores tend to scatter about the mean of a set of data. If the standard deviation is relatively small, a close clustering of scores about the mean is indicated. A relatively large value indicates wide scattering about the mean.

In the factory data of Table 16, the standard deviation would be much greater for Factory A than for Factory C. The standard deviation for Factory B would be greater than that of Factory D, even though their ranges are the same.

The small Greek s (sigma), σ, is used to represent the standard deviation, and the complete standard deviation formula looks like this:

$$\text{Standard deviation} = \sigma = \sqrt{\frac{\Sigma(X - M)^2}{N}}$$

where X is a score
            M is the mean
            N is the number of scores
            Σ means "the sum of all"

Let's see how this formula works in finding the standard deviation for these scores: 11, 12, 13, 14, 15, 16, 17.

First we find the mean.

$$M = \frac{\Sigma X}{N} = \frac{11 + 12 + 13 + 14 + 15 + 16 + 17}{7} = \frac{98}{7} = 14$$

Then we construct Table 17 to find the sum of the squares of the deviations from the mean.

Table 17

Score (X)	Deviation from the mean (X — M)	Deviation from the mean squared (X — M)²
17	3	9
16	2	4
15	1	1
14	0	0
13	— 1	1
12	— 2	4
11	— 3	9

$$\Sigma(X - M)^2 = 28$$

Then we compute the standard deviation.

$$\text{Standard deviation} = \sigma = \sqrt{\frac{\Sigma(X - M)^2}{N}} = \sqrt{\frac{28}{7}} = \sqrt{4} = 2$$

The standard deviation is a number that we can use for comparison. How does your weight compare with the weight of other persons of your age? Suppose that you are 10 pounds heavier than the average boy of your age. Are there many other boys of your age heavier than this, or are you a little giant? You need to know the standard deviation of weights of boys your age before you can know much about how you compare with others in weight. Suppose the mean weight for boys of your age is 100 pounds, and the standard deviation is 5 pounds. Then your weight of 110 pounds is as much as two standard deviations, (2×5) pounds, above the mean. After you have learned more about statistics, you will find that this means that, in general, only two persons in 100 of your age are heavier than you.

Let's use the standard deviation to study some more test scores. Suppose that the mean score on a standard science unit test is 42 and the standard deviation is six points. If you have a score of 48, you are above the mean. You are six points above the mean, or as many points as one standard deviation above the mean. Suppose that the mean on a different unit test is 25 and the standard deviation is 10. If you had a score of 35 on this test, you would be 10 points, or one standard deviation, above the mean. Then you would know that you did equally well (with respect to the other class members) on both tests. You are one standard deviation above the mean on each test. Suppose Kim had a score of 30 on the first test. His score was 12 points, or two standard deviations, below the mean. What would his score on the second unit test be if he did just as poorly on it as on the first unit test? Jane had a score of 45 on the first test and 35 on the second test. On which test did Jane do better?

EXERCISE SET 13
Measures of Dispersion

1. Compute the standard deviations for these scores:
    a. 1, 2, 3, 4, 5, 6, 7
    b. 6, 7, 8, 9, 10, 11, 12, 13, 14

2. On a spelling test at Pillsbury School, the mean was 54 and the standard deviation was 8. Copy and complete this table by changing each test score to the number of standard deviations from the mean.

Pupil	Test score	Points from the mean	Number of standard deviations from the mean
Mark	62	$62 - 54 = 8$	$\frac{8}{8} = 1$
Rod	46	$46 - 54 = -8$	$-\frac{8}{8} = -1$
Sue	70		
Kay	50		
Charles	54		
Jane	66		

3. On another spelling test at the same school two weeks later, the mean was 28 and the standard deviation was 10. Copy and fill in the blanks in this table.

Pupil	Test score	Points from the mean	Number of standard deviations from the mean
Mark	38		
Rod	28		
Sue	43		
Kay	18		
Charles	33		
Jane	23		

    a. Which pupil improved most in regard to standard deviations?

b. Which pupil came down the most on the second test?

c. Which pupil had a similar score on each test?

4. Copy and complete the following frequency distribution table of test scores on a ten-point test and compute the mean and standard deviation for the data.

Score (X)	Frequency (f)	fX	X — M.	(X — M)²	f(X — M)²
0	1	0			
1	0				
2	1				
3	1				
4	3				
5	3				
6	5				
7	8				
8	2				
9	2				
10	1				

5. Write a formula for the standard deviation of a frequency distribution in terms of f, X, M, and N.

Probabilities from Measures of Scatter

The senior class of Central High School was sponsoring an all-school carnival. One of the boys in the class knew the operator of an amusement park, and obtained from him a bag containing over a thousand circular disks, each marked with a number from 1 to 7. The class planned to give away prizes by letting people draw a number from the bag, but they needed to know the distribution of numbers in the bag in order to determine which number deserved the best prize. They were unable to obtain this information and they didn't want to examine each disk, so they decided to conduct a statistical experiment to get an idea of the distribution of the disk numbers. They started by drawing sixty-four disks. The results of the draw are summarized in Table 18.

Table 18

Mark	Frequency (number of disks)
1	1
2	6
3	15
4	20
5	15
6	6
7	1

The table indicates that one disk had a mark of 1, six disks had a mark of 2, and so on.

The class found the mean score to be 4 and the standard deviation to be approximately 1.2. Check these results with your own computation.

If the class assumed that this draw was a good representation of the distribution of disks in the bag, they could use these results to estimate the probability of drawing certain numbers from the bag. From their data, there are sixty-four ways to draw a number but only one chance for a 7, so the probability of drawing a 7 might be estimated as $\frac{1}{64}$ or .016. This would also be the probability for drawing a 1.

The probability for drawing a 6 *or* a 7 could be estimated in this way: since there are sixty-four possible ways to draw indicated, one possible way to draw a 7, and six possibilities for a 6, the probability is $\frac{7}{64}$ or .11.

A histogram of the data of Table 18 is shown in Figure 7. A frequency polygon has been drawn in Figure 7 by joining the midpoints of the histogram bars, and the frequency polygon has been changed to a smooth curve as indicated by the broken line in the figure. Notice that this smooth curve is approximately bell-shaped. A bell-shaped graph indicates many scores near the mean of the data, with the number of scores gradually decreasing as you go away from the mean so that there are a few low scores and a few high ones.

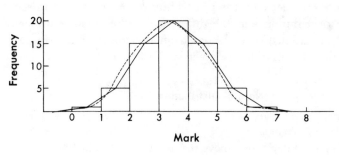

Figure 7

A perfect bell-shaped curve that represents the distribution of an endless set of data is called a *normal curve*. The picturing of data by a normal curve is called a *normal distribution*.

If the seniors at Central High assume that the distribution of numbers in their bag of disks is the same as that in their sample, it would be close to (but not exactly the same as) a normal distribution.

It has been found that many distributions of data closely follow the pattern of a normal distribution. For example, if you were to collect data on the heights of thousands of six-year-old children, on the weights of thousands of apples, on the thicknesses of thousands of metal parts made by an automotive company, or on the scores of thousands of students on a standard mathematics test, you would probably obtain distributions that would be approximately like a normal distribution. Of course, you must also remember that not all data distributions follow the pattern of a normal distribution.

There are many important ideas related to normal distributions and normal curves, but many of them are too complex to discuss here. However, we can learn some important ideas related to a normal distribution by examining the numbered-disk example presented earlier in this section. In that example, we saw that the greater the difference of a score from the mean, the lower the probability of getting that score. For example, the probability of drawing a 7 was only $\frac{1}{64}$ (or, in other words, there was about a 2% chance of getting a 7), while the probability of getting a 6 or a 7 was about $\frac{7}{64}$ (about an 11% chance). Mathematicians have found that in distributions that are approximately normal, there is a relationship between the probability of getting a certain score from the distribution and the difference, expressed in number of standard deviations, of the score from the mean. These findings for a normal distribution are given in Table 19.

Table 19
Normal Distribution Table

Number of standard deviations from the mean	How often scores this high or higher are expected
2 standard deviations above the mean	2 in 100
$1\frac{1}{2}$ standard deviations above the mean	7 in 100
1 standard deviation above the mean	16 in 100
$\frac{1}{2}$ standard deviation above the mean	31 in 100
The mean (0 standard deviations from the mean)	50 in 100
$\frac{1}{2}$ standard deviation below the mean	69 in 100
1 standard deviation below the mean	84 in 100
$1\frac{1}{2}$ standard deviations below the mean	93 in 100
2 standard deviations below the mean	98 in 100

The first line of this table says that the probability of a score being two standard deviations or more above the mean of a normal or nearly normal distribution is .02. The second line says that the probability of getting a score one and one-half standard deviations or more above the mean is .07, and so on. The results from the numbered-disk example are similar to the figures presented in Table 19.

If you know that a distribution of data is approximately normal and know the mean and standard deviation for the data, you can use Table 19 to determine the probability of getting a certain score from the data. Going back to an example involving weights, if we assume that weights of children for a certain age group form a normal distribution, then a weight two standard deviations or more above the mean weight would be likely to occur only two times out of one hundred. In other words, if your weight is two standard deviations above the mean weight for your age group, it is probable that only 2% of the people in your age group are as heavy as or heavier than you. Another way to interpret this is to say that your weight would place you in the 98th-percentile weight rank for your age group. A percentile-rank interpretation for a normal distribution is given in Table 20.

Table 20

Normal Distribution Table

Number of standard deviations from the mean	Percentile rank in a normal distribution
+ 2.0 (two standard deviations *above* the mean)	98
+ 1.5 ($1\frac{1}{2}$ standard deviations above the mean)	93
+ 1.0 (one standard deviation above the mean)	84
+ .5 ($\frac{1}{2}$ standard deviation above the mean)	69
0 (the average or mean score)	50
− .5 ($\frac{1}{2}$ standard deviation *below* the mean)	31
− 1.0 (one standard deviation below the mean)	16
− 1.5 ($1\frac{1}{2}$ standard deviations below the mean)	7
− 2.0 (two standard deviations below the mean)	2

If the mean score on a nation-wide test is 42, if the standard deviation is 6, and if the scores form a normal distribution, a score of 48 on the test would give a percentile rank of 84. That is, it is *probable* that 84% of the students taking this test would rank below a person having a score of 48, and about 16% would rank above that person.

Note the similarity between Table 19 and Table 20. Table 19 indicates that the chance of a score of 48 or higher is $\frac{16}{100}$ or .16, and the chance of a score lower than 48 is $\frac{84}{100}$ or .84.

EXERCISE SET 14

Chance and the Normal Distribution

1. From the data given in problems 2 and 3 of Exercise Set 13, find the percentile rank of the Pillsbury pupils on the spelling tests. (Assume that all spelling test scores would form a normal distribution. Use Table 20.)

Pupil	Percentile rank on first test	Percentile rank on second test
Mark		
Rod		
Sue		
Kay		
Charles		
Jane		

2. Use Table 19 to find out how often we expect the boys of a certain grade to have certain weights. Assume a normal distribution with a mean of 100 and a standard deviation of 5. Copy the table below and fill in the blanks.

Weight	Pounds from the mean	Standard deviations from the mean	How often expected to be this weight or heavier
100	$100 - 100 = 0$	0	50 in 100
105			
110			
95			
90			
$102\frac{1}{2}$			
$92\frac{1}{2}$			

3. When we compare the measurements of two groups, we can use the normal distribution tables to tell us how often we expect measures to differ in certain amounts. Suppose that the mean difference in two measures is 8 and the standard deviation of this difference is 2. If we assume these differences form a normal distribution, how often do we expect to get the differences in the table below? Copy the table below and fill in the blanks.

Difference in Measurements

Difference in two measures	Points from the mean difference	Standard deviations from the mean	How often these differences are expected
10	$10 - 8 = 2$	$\frac{2}{2} = 1$	16 in 100
6			
8			
9			
5			
12			

This is very much like the way we use the standard deviation to study samples. The normal-curve table tells us how often we expect to get certain scores with samples.

4. Observe the following number pattern.

$$1$$
$$1 \quad 1$$
$$1 \quad 2 \quad 1$$
$$1 \quad 3 \quad 3 \quad 1$$
$$1 \quad 4 \quad 6 \quad 4 \quad 1$$
$$1 \quad 5 \quad 10 \quad 10 \quad 5 \quad 1$$

Notice that each row starts and ends with 1 and that the other numbers are equal to the sum of the two numbers directly above them.

    a. Complete three more rows of this pattern.

    b. Use the numbers in your last row as frequencies of a frequency distribution, and construct a histogram for the distribution.

    c. Do you think your distribution is something like a normal distribution? As you increase the number of rows, what do you suspect will be the relationship between the numbers in the last row and the frequencies of a normal distribution?

5. The maximum daily temperatures for January for a certain city form a distribution that is approximately normal. The mean temperature of this distribution is 38° Fahrenheit and the standard deviation is 6° F.

    a. What is the probability of having a maximum temperature of 47° F. or greater in this city in January?

    b. What is the probability of having a maximum temperature of 26° F. or lower in this city in January?

    c. What is the probability of having a maximum temperature between 32° F. and 44° F. in this city in January?

How Good Is the Prediction?

We have examined the desirability, necessity, and usefulness of working with samples of numerical information. We have also seen some examples of how probability ideas can be used to answer statistical questions. Some of the most important tasks of statisticians involve both sampling and probability. In the preceding sections of Part VI, we have seen examples of using samples, probability, and the normal distribution to make predictions. But these examples bring forth other questions. What is the chance that a sample truly represents the group from which it was selected?

If we make a prediction from a sample, what is the probability that the prediction is correct? In other words, the statistician is often concerned with the question, "How good is the prediction?"

When attempting to make conclusions about a group of things or quantities by using a random sample selected from the group, the statistician often proceeds in one of two ways: (1) He might make some guess about a statistical measure of the group and then try to decide how good his guess was by working with a sample from the group; (2) He might start by taking a sample and compute some statistical measure of the sample such as the mean. Then he might give an estimate of the mean for the entire group that would have a high probability of being correct.

Let's look at an example regarding the control of the quality of a factory product. Suppose that a ball-point pen manufacturer wants to study the quality of his pens. A typical question he might ask is, "How many defective pens are the machines making?"

In answering this question with method (1), he might guess that one pen out of every ninety will be defective, and select that as a reasonable average for proper production. He might then check this guess by testing samples of the pens (he would not want to test every pen) and find that the samples average one defective pen in every eighty. Statistical methods can tell if this result is close enough to justify acceptance of the 1-out-of-90 estimate.

On the other hand, he might want to use method (2) to predict a certain range for the ratio of defective pens. To do this, he would collect samples, compute the average ratio of defective pens for the samples. He would then use statistical methods to estimate a normal range for the ratio of defective pens.

Of course, in both situations the manufacturer would have the problem of how to properly select samples that would give a lot of confidence in the results.

Once he had a statistical estimate of the ratio of defective pens produced, he might use statistical methods to answer other questions, such as "Is the ratio of defective pens one that will enable the company to maintain a reputation for a product of high quality?"

We cannot develop, in this book, the various methods of statistics that have just been mentioned. However, this gives you an idea of the interesting and important ideas to which a study of statistics can lead.

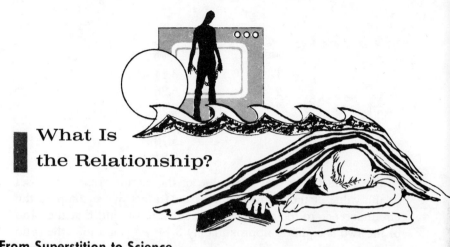

What Is
the Relationship?

From Superstition to Science

Is there any relationship between the amount of sleep you get and your achievement in mathematics? Do TV shows cause an increase in crime? Statistics are used to find answers to questions like these.

So far, we have been concerned with such things as averages, deviations, and predictions from sets of data. Now we take up the important problem of finding relationships *between* sets of data.

Most of the discoveries in science have been the result of finding relationships between data. A typical illustration is the relationship between ocean tides and the positions of the sun and the moon. As soon as scientists found this relationship, they knew what caused our ocean tides. And as soon as they knew the cause, they could predict the times and sizes of the tides.

But events sometimes occur together by chance and are falsely thought to be related. A false relationship such as an unfortunate experience and the presence of a black cat leads to a belief in superstitions. An old superstition from the Indians is the idea that the phases of the moon are related to weather conditions. We must be careful not to consider events that happen at the same time to have a cause-and-effect relationship. By this we mean that, when two things happen at the same time, one of the events need not be the cause of the other.

In mathematics, we often find it necessary to make comparisons between sets of numbers. For example, Table 21 shows a comparison between the lengths of the diameters of certain circles and the circumferences of the circles.

Table 21

Diameter	Circumference
1 in.	3.14 in.
2 "	6.28 "
3 "	9.42 "
4 "	12.56 "
5 "	15.70 "
6 "	18.84 "

The table shows that the value of the circumference increases as the value of the diameter increases. In fact, if we compare the circumference to the diameter in each case, we find that the ratio of the two is equal to approximately 3.14. As you know, the ratio of the circumference of any circle to its diameter will always give the same value, and we use the symbol π to represent that constant value. We express the relationship between the diameters of circles and their circumferences with the formula $C \div d = \pi$ or $C = \pi d$. In Figure 8, a graph is used to picture the relationship between the corresponding pairs of numbers listed in Table 21. We see that the points that represent the corresponding circumference and diameter values lie on a straight line. The formula and the graph indicate that the diameter and the circumference of a circle *change at the same rate*. We often describe such a relationship as being *perfect* or *exact*.

Comparing the Diameters and Circumferences of Circles

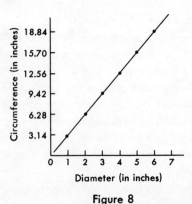

Figure 8

Table 22 shows a comparison between the weights of eight boys and their scores on an English test.

Table 22

Weight	Test score
90	22
100	20
105	26
110	15
115	8
120	25
125	28
130	10

The pairs of values obtained from Table 22 are pictured in the graph of Figure 9. The graph and the table indicate no apparent relationship between the two sets of data. It is impossible to describe this relationship with a formula.

Comparing Weights and Test Scores for Eight Boys

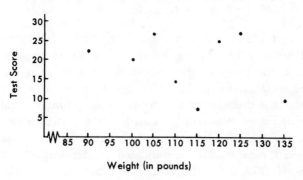

Figure 9

Table 23 shows a comparison between the number of cups of hot chocolate sold at football games during a season and the temperature the night of each game.

Table 23

Temperature	Number of cups
70° F.	20
65	24
55	34
50	38
40	50
30	64

This data is pictured in Figure 10.

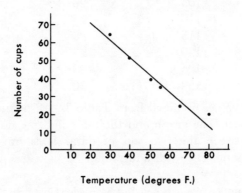

Comparison Between Temperature and
Number of Cups of Hot Chocolate Sold

Figure 10

The graph and the table indicate some relationship between the temperature and the number of cups of hot chocolate sold, for we see that, as the temperature decreased, the number of cups sold increased. A relationship in which an increase in one quantity brings about a decrease in a corresponding quantity is called an opposite or *inverse* relationship. This is not a perfect inverse relationship, for the ratio of the two quantities being compared is not always the same. This is indicated on the graph by the fact that the points of the graph do not lie exactly on a straight line. Again, we would not be able to find a formula that would exactly describe this relationship.

We have examined three situations in which sets of data have been compared. In each case, we attempted to describe the comparison in words. Although it is not always possible to describe a relationship with a mathematical formula, in statistics it is possible to numerically describe the relationship between two sets of data. The methods used to do this are too difficult to present here, but we can gain an idea of the end result of the methods. A relationship between sets of data is called a *correlation*.

The *amount* of relationship between two sets of data is measured by what is called the *correlation coefficient*. Correlation coefficients have values from 0 to 1. No relationship is indicated by a correlation coefficient of 0. A perfect relationship is indicated by 1. When a relationship is inverse, we use negative numbers. A perfect inverse relationship would be described by − 1. The relationship between the circumference and diameter of circles is 1. The correlation between weights and test scores in the second example is 0. The correlation coefficient for the third example might be − .8 or − .9. The computation would be a bit involved, however.

EXERCISE SET 15
Working with Data Relationships

1. Suppose that a correlation is computed for the following quantities. Mark with a + sign the correlations which you expect to be positive and with a − sign those which you expect to be inverse or negative. A 0 may be used to indicate no relationship.

   a. Rainfall and attendance at baseball games.
   b. The age of a car and its trade-in value.
   c. The price of corn and the price of beef.
   d. Scholarship and success in life.
   e. Amount of unemployment and wage rates.
   f. Temperature in Minnesota and tourists in Florida.
   g. Cost of life insurance and age at which an adult buys it.
   h. Beauty and intelligence.
   i. Total population and total school enrollment.
   j. Area of a circle and its radius.
   k. Length and width in the formula LW = 12.
   l. Amount of smoking and incidence of lung cancer.
   m. Miles driven and amount of gasoline consumed.

2. Give three illustrations of positive correlations that have not been listed in Part VI.

3. Give three illustrations of negative correlations that have not been listed in Part VI.

4. Positive correlation coefficients have been found between the following measures. It is obvious that no cause-and-effect relationship exists. Give a probable reason why the correlation coefficient is high.

   a. Number of inmates in institutions for the insane and the number of automobile accidents.
   b. Number of TV sets in the United States and the average salary of college professors.
   c. The consumption of tobacco and the decrease in deaths from polio.

Using Your Knowledge

Data for Investigation

If you want to collect data about current events, the list below will suggest some possible explorations. You can then use these data to apply the principles discussed in this part of the book. Label completely the data you collect, describe the method of collection, draw a graph of the data, and write questions that can be answered from the data.

1. Scoring records at athletic events
2. Weather reports — temperatures, rainfall, humidity, storms
3. Traffic records — accidents, amount of traffic, number of vehicles, number of parking places
4. School absences or tardinesses
5. School costs
6. Lunchroom or candy sales
7. Vital statistics — births, deaths, marriages, unemployment
8. Recreational activities of friends — radio, movies, books, magazines
9. Market quotations — stocks, grains, cattle
10. Newspapers — ads, pictures, comics
11. Business conditions — sales, prices, bank deposits, interest rates
12. Tax rate and expenditures
13. School grades and marks

14. Money in circulation, public debt
15. Electrocardiograms, temperature readings, respiration rates
16. Utility bills — gas, electric, water
17. Use of letters, words, numbers per page in a book
18. Clothes inventory — color, number, type
19. Heights, weights, shoe sizes
20. Distribution of birthdays, and the probability of some birthday falling within a certain time period
21. Coin dates

Here are some more activities that will enable you to apply many of the ideas found in Part VI.

1. Perform an experiment to test the laws of chance. Suggested materials: coins, dice, cards, mortality tables, bingo numbers, mind reading, slot machines.
2. Make a bulletin-board exhibit or chart of examples of statistics in daily life.
3. Make a model, chart, or device to illustrate a normal distribution.
4. Visit an insurance actuarial office or statistics laboratory or interview a statistician.

A Backward Glance and a Look to the Future

Whenever you see a news report, an advertisement, or a new discovery that involves numerical data, test the result with these questions:

Who says so? Does the person reporting have a prejudice? Is he qualified to state the answer?

How does he know? Has he used a random sample that represents all points of view? How did he get the sample? How many and how large were the samples? Is the relationship reasonable?

What's missing? Is the bias for comparison given? What is the probability that the statement is true?

Now that you have read this section and have worked some exercises, you should be able to understand many things in the world of statistics. You should be able to read tables of data and know what they are about. You should not be fooled by graphs that try to create a false idea. You should understand the meanings of claims made in advertisements. You should be able to understand and compute such things as means, medians, percentile ranks, and standard deviations. You should have some ideas about the use of samples and probability in discovering new facts. You should enjoy your reading more, because you know how questions are answered by the use of statistics. The world will be more interesting because you know something about the world of statistics.

New applications for statistical methods are constantly being discovered. The use of modern electronic computers for calculations has enabled mathematicians and scientists to solve statistical problems that once were considered impossible. Applications of statistical methods will continue to expand, and statistics will play an important part in the space age of the future. If you should decide to become a statistician, you have an interesting career ahead of you.

Extending Your Knowledge

BEARDSLEY, MONROE, *Thinking Straight*. Prentice-Hall, Inc., 1956

BROSS, IRWIN D., *Design for Decision*. Macmillan Co., 1953

COMMISSION ON MATHEMATICS, *Introductory Probability and Statistical Inference for Secondary Schools*. College Entrance Examination Board, 1957

DIXON, W. J., and MASSEY, F. J., JR., *Introduction to Statistical Analysis*. (Difficult but complete.) McGraw-Hill, 1951

HUFF, DARRELL, *How to Lie with Statistics*. Norton & Co., 1954

HUFF, DARRELL, and GEIS, IRVING, *How to Take a Chance*. Norton and Co., 1959

KENENY, JOHN G., SNELL, J. LAURIE, and THOMPSON, GERALD L., *Introduction to Finite Mathematics*. Prentice-Hall, Inc., 1957

MORONEY, M. S., *Facts from Figures*. Penguin Books, 1956

NATIONAL INDUSTRIAL CONFERENCE BOARD, *Weekly Graphs* and *Economic Almanac*. 460 Park Avenue, New York 22, New York

NATIONAL COUNCIL OF TEACHERS OF MATHEMATICS, *The Growth of Mathematical Concepts*. Twenty-fourth Yearbook, 1959

WALLIS, W. ALLEN, and ROBERTS, HARRY V., *Statistics: A New Approach*. The Free Press, 1956

Solutions to the Exercises

Solutions to the Exercises PART I

EXERCISE SET 1

1. They are never free on the same day.

2. a. HHH — TTT b. HH — HTTT c. HHTH — TT

d. HHTHT — T e. HHT — THT f. H — THTHT

g. — HTHTHT h. TH — HTHT i. THTH — HT

j. THTHTH — k. THTHT — H l. THT — THH

m. T — THTHH n. TT — HTHH o. TTTH — HH

p. TTT — HHH

The solution depends on the pattern of alternating HTHT wherever possible.

EXERCISE SET 2

1. A number is divisible by 9 if the sum of the digits is divisible by 9; hence, 477, 648, and 8766 are divisible by 9.

2. The square of an odd number is odd. The square of an even number is even. The square of a number divisible by 5 is also divisible by 5. The square of 22 will be divisible by 2 and 11.

3. The quotients will all be about $3\frac{1}{7}$, or equal to π.

4. As the pendulum increases in length, the time of its swing increases.

EXERCISE SET 3

1. 3 minutes.

2. One coin is a half-dollar: the *other* one is a nickel.

3. The cork costs $2\frac{1}{2}$ cents.

4. 28 days.

EXERCISE SET 4

1. *a.* Inductive. *b.* Deductive. *c.* Inductive.

2. Every native would say, "I am a Bau Wau." Hence, the second is a Bau Wau and the third is a Mau Mau.

3. The men paid $27, of which the hotel got $25 and the bellhop $2.

4. Suppose the candidates are A, B, and C. A reasons that if he had a blue cross, then B or C would immediately know that they had a red cross. Since B and C could not determine their color, A must also be marked with a red cross. This is an example of indirect reasoning.

5. The grocer lost $7 and $3 worth of groceries.

6. It can't be done. Each domino covers a black and a red square. Since 2 black squares (or 2 red squares) are cut off, there are 2 extra red squares left.

EXERCISE SET 5

1. $2+3 = 2+(2+1)$
$= (2+2)+1$
$= 4+1$
$= 5$

2. $2\times2 = 2(1+1)$
$= 2\times1+2\times1$
$= 2+2$
$= 4$

1. 59

2. $\begin{array}{r} 9\,5\,6\,7 \\ +\,1\,0\,8\,5 \\ \hline 1\,0\,6\,5\,2 \end{array}$

3.

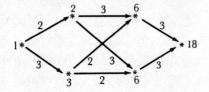

4. Give each puppy a number, such as P_1, P_2, etc.
   a. Balance P_1 P_2 with P_3 P_4.
   b. If they balance, compare P_5 P_6 on the scale.
   c. If the weighing in a does not balance, compare the two heavy puppies on the scale.

EXERCISE SET 7

1. a. 3 b. 0 c. 1 d. 1 e. 4 f. 0 — all mod five
2. Yes.
3.

+	0	1	2	3	4
0	0	1	2	3	4
1	1	2	3	4	0
2	2	3	4	0	1
3	3	4	0	1	2
4	4	0	1	2	3

4. a. No b. No c. No
5.

×	0	1	2	3	4
0	0	0	0	0	0
1	0	1	2	3	4
2	0	2	4	1	3
3	0	3	1	4	2
4	0	4	3	2	1

6. a. 1 b. 4 c. 3 d. 0
   e. 2 f. 3 — all mod five
7. a. 2 b. 2 c. 3 d. 3
   e. 2 f. 4 — all mod five
8.

+	0	1	2	3
0	0	1	2	3
1	1	2	3	0
2	2	3	0	1
3	3	0	1	2

×	0	1	2	3
0	0	0	0	0
1	0	1	2	3
2	0	2	0	2
3	0	3	2	1

EXERCISE SET 8

1. At the North Pole or about $1\frac{1}{8}$ miles from the South Pole.
2. B is opposite Y, R is opposite G, and P is opposite W.

EXERCISE SET 9

1.

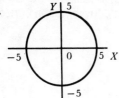

2. $S = 16t^2$

EXERCISE SET 10

1. The dice can fall 36 different ways. Of these, (6, 1), (5, 2), (4, 3), (3, 4), (2, 5), (1, 6), or 6 out of 36, total 7.

2. Draw from the BW box. If you draw a black, then the box has 2 blacks, the WW box has a black and white, and the BB has 2 white ties. If you draw a white from the BW box, then this box has 2 white ties, the BB box has a black and white tie, and the WW box has 2 black ties.

EXERCISE SET 11

1. e

2. Each group is made up of special students who may not be representative of all students in your school.

3. *a.* Floating is related to cleansing qualities.

b. Low cost is the greatest factor in determining the quality of a bicycle.

c. Eating Bumper cereal is related to athletic skill.

EXERCISE SET 12

1. 12:58

3. An infinite number.

2. 2

4. Zero.

EXERCISE SET 13

1. *a.* Poorer.　　*b.* Poorer.　　*c.* Richer.　　*d.* Richer.

e. Richer.

Mr. Robinson	$5
Mr. Carr	$1255
Mr. Brown	$2505
Mr. Smith	$3755
Mr. Jones	$3755
Mr. Stevens	$5005

2. The girl will marry John because he has more of the required qualities than any other man. She will marry none of them if she insists that the man have all of the qualities she is looking for.

EXERCISE SET 14

1. Every number can be written as the sum of 4 squares or less.

2. 8 cubes have 3 sides painted.

　12 cubes have 2 sides painted.

　6 cubes have 1 side painted.

　1 cube has no side painted.

1. There is no formula that will give the prime numbers.
2. a. (3, 5), (5, 7), (11, 13), (17, 19), (29, 31), (41, 43), (59, 61), (71, 73)
   b. 0, 4, 4, 10, 10, 16, 10
   c. No.
3. a. 5+3   b. 19+7   c. 11+7   d. 37+11
4. a. Yes.   b. Yes.   c. Yes.
5. a. No.   b. No.

Solutions to the Exercises PART II

Exercise Set 1

1. *a.* discrete   *b.* continuous
   c. discrete   *d.* discrete
   e. continuous   *f.* continuous
2. *a.* approximate *b.* approximate
   c. approximate *d.* approximate
   e. exact   *f.* exact
3. *a.* 2   *b.* 24
   c. 20   *d.* 4
   e. 2   *f.* 144   *g.* 500

Exercise Set 2

1. *a.* $1\frac{1}{2}$ miles   *b.* 6 feet

For answers *c* and *d*: a cubit is equal to approximately 18 inches.

   c. 450 ft. by 75 ft. by 45 ft.
   d. 90 ft. by 30 ft. plus a porch 15 ft. by
        30 ft. (I Kings 6: 2-3)
2. Measures depend on person selected.
3. About 18 inches
4. About 5 feet

Exercise Set 3

1. Place any three coins in the left pan and any other three in the right pan. If they balance, the counterfeit is one of the two remaining coins, and the second weighing will disclose the false coin. If the pans do not balance, the counterfeit is in the group of three coins that shows light. Place a coin of this group of three in each pan. If the scales balance, the remaining coin is false; if the scales do not balance, the counterfeit is the light coin.
2. 1, 2, 4, 8, 16

Exercise Set 4

1. *a.* $A = \frac{1}{2} bh$ *b.* $A = r^2$ *c.* $A = bh$

2. *a.* 6   *b.* 9

3. *a.* Usually $31\frac{1}{2}$ U.S. gallons or 36 imperial gallons.

   b. 63 to 140 gallons, depending upon the commodity being measured.

   c. Usually $\frac{1}{4}$ barrel but sometimes measured in pounds.

355

4. 4620 cubic inches

5. dry pint = 33.6 cu. in.; liquid pint = 28.9 cu. in.; dry pint is 4.7 cu. in. larger.

6. British gallon = $1\frac{1}{5}$ U. S. gallon

7. 4 pecks = 1 bushel,   8 quarts = 1 peck

8. 27

Exercise Set 5

1. 10 cm.   **2.** 3.8 cm.   **3.** 50 m.

4. 200 sq. cm.   **5.** 2,000,000 cc.   **6.** 20.32 cm.

7. 38.4 km. (using 1 km. = $\frac{5}{8}$ mile)   **8.** 13.2 pounds   **9.** 4.228 qt.

10. 4.5 kg.   **11.** $7\frac{1}{2}$ miles (Using 1 km. = $\frac{5}{8}$ miles)

Exercise Set 6

1. Gregorian

2. About 39 inches

3. $\frac{1}{8}$ sec. $\frac{1}{4}$ sec. $\frac{1}{2}$ sec.   1 sec.   2 sec.

Exercise Set 7

1. 57°   **2.** 60°

Exercise Set 8

1. 21 ft.-lb.   **3.** 60   **4.** .0000082 mm.

5. *a.* England   *b.* England   *c.* England   *d.* Mexico
e. France   *f.* Italy   *g.* Germany   *h.* Russia
Check market quotations for current value.

6. mite = about $\frac{1}{3}$ cent

silver shekel = 50 to 60 cents,   gold shekel = about $10
talent = 3000 shekels

Exercise Set 9

Answers depend on local measurements

Exercise Set 10

1. *a.* counted   *b.* counted   *c.* approximate
d. approximate   *e.* counted   *f.* counted

2. *a.* indirect   *b.* direct   *c.* indirect
d. indirect   *e.* indirect   *f.* direct

Exercise Set 11

1. *a.* 1 min. $\frac{1}{2}$ min.

   b. $\frac{1}{8}$ in. $\frac{1}{16}$ in.

   c. .01 ft. .005 ft.

   d. $\frac{1}{16}$ in. $\frac{1}{32}$ in.

   e. $\frac{1}{16}$ cu. ft., $\frac{1}{32}$ cu. ft.

2. *a.* 7.45 in. 7.35 in.

   b. $16\frac{7}{8}$ in. $16\frac{5}{8}$ in.

   c. 6.935 cm. 6.925 cm.

   d. $3\frac{1}{16}$ in. $2\frac{15}{16}$ in.

Exercise Set 12

1. *a.* $\frac{1}{2}$ ft. *b.* .05 lb. *c.* $\frac{1}{4}$ min.

   d. .005 gm. *e.* 5 miles *f.* .0005 m.

2. *a.* .083 *b.* .010 *c.* .009

   d. .125 *e.* .009 *f.* .00006

3. *a.* .1   .05   .0089

   b. 10   5   .0089

   c. .001   .0005   .0089

   d. 100   50   .0089

4. 0.056 is most precise.
   All have the same accuracy.

5. *a.* 56×0.1

   b. 56×10

   c. $56 \times .001$

   d. 56×100

Exercise Set 13

1. *a.* 572   3

   b. 68   2

   c. 1370   4

   d. 9   1

2. *a.* 3   *b.* 2   *c.* 3   *d.* 3   *e.* 2   *f.* 4   *g.* 5

Exercise Set 14

1. 8460   2. .74

3. 474,000   4. 84

5. 245,100   6. 406

Exercise Set 15

1. *a.* 8.8 sec.   *b.* $8\frac{0}{2}$ in.   *c.* 8 lb. 0 oz.

d. $10\frac{0}{2}$ in.   *e.* 15 gm.   *f.* 26 cm.

2. *a.* 3.0 liters   *b.* $8\frac{0}{2}$ in.   *c.* 1 gal. 3 qt.

d. $5\frac{1}{2}$ ft.   *e.* 1 gm.   *f.* 4 m. 22 cm.

Exercise Set 16

1. *a.* 9.2 cm.   *b.* 200 sq. in.   *c.* .4 sq. ft.   *d.* .06 cu. in.
2. *a.* 2 ft. 4 in.   *b.* 30   *c.* 1   *d.* 13

Exercise Set 17

1. *a.* $2\frac{1}{2}$ *b.* 175   *c.* 19   *d.* 1800

2. *a.* 13 rays 1 say   *b.* 8 rays 15 says
   c. 46 rays 10 says   *d.* 6 says 16 ways
3. *a.* 8 dills 1 pill   *b.* 4 dills
   c. 46 sq. mills   *d.* 1 dill 2 pills

Exercise Set 18

1. *a.* $3.05 < \pi < 3.15$ or $\pi \approx 3.1$
   b. $0.66 < \frac{2}{3} < 0.67$ or $\frac{2}{3} \approx .67$
   c. $1.5 < x+y < 2.5$ or $x+y \approx 2$
2. *a.*

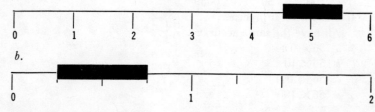

   b.

Exercise Set 19

1. *a.* .1 in.   *b.* .01 ft.   *c.* $\frac{1}{8}$ in.   *d.* 1 oz.   *e.* .001 sec.

2. *a.* $\frac{1}{2}$ min.   *b.* .05 oz.   *c.* $\frac{1}{8}$ in.   *d.* 500 miles   *e.* .005 ft.

3. *a.* 6.94   *b.* 10.60   *c.* .01   *d.* 56.00   *e.* 4.00
4. *a.* 468   *b.* .00708   *c.* 3.47   *d.* 5690   *e.* 3.00
5. d, a, b, c, e
6. d, e, b, a, c

7. *a.* 4   *b.* 4   *c.* 4   *d.* 3   *e.* 2   *f.* 1

8. *a.* 20.8   *b.* $8\frac{0}{2}$ *c.* 5.9   *d.* $7\frac{0}{16}$

9. *a.* 27   *b.* 4.7   *c.* 40   *d.* .6

10. *a.* .57   *b.* 2.2   *c.* 30   *d.* 9.000

11. *a.* F   *b.* F   *c.* F   *d.* T   *e.* F

12. *a.* .1 in.   .05 in.   .0016   3

     b. .01 sec.   .005 sec.   .0006   3

     c. $\frac{1}{8}$ mile $\frac{1}{16}$ mile   .0067   2

     d. 10 lb.   5 lb.   .0625   1

     e. .1 qt.   .05 qt.   .0028   3

     f. 1000 cu. ft.   500 cu. ft.   .1000   1

     g. .0001 mm.   .00005 mm.   .0100   2

     h. .001 cc.   .0005 cc.   .0005   4

13. 580,000,000 miles per year

       1,600,000 miles per day

         67,000 miles per hour

Solutions to the Exercises

PART III

EXERCISE SET 1

1. • 1,4 • 2,4 • 3,4 • 4,4
   • 1,3 • 2,3 • 3,3 • 4,3
   • 1,2 • 2,2 • 3,2 • 4,2
   • 1,1 • 2,1 • 3,1 • 4,1

2. *a.* •   •   •   •   *b.* •   •   •   •   *c.* ⊙1,4 ⊙2,4 ⊙3,4 •
      •   •   • ⊙4,3   •   •   •   •   ⊙1,3 ⊙2,3 •   •
      •   • ⊙3,2 •   •   •   •   •   ⊙4,2 ⊙1,2 •   •   •
      • ⊙2,1 •   •   • ⊙2,1 •   •   •   •   •   •

3. • −4,4   • −2,4   • 0,4   • 2,4   • 4,4
   • −4,2   • −2,2   • 0,2   • 2,2   • 4,2
   • −4,0   • −2,0   • 0,0   • 2,0   • 4,0
   • −4,−2 • −2,−2 • 0,−2 • 2,−2 • 4,−2
   • −4,−4 • −2,−4 • 0,−4 • 2,−4 • 4,−4

4. *a.* No points are circled.

   b.   4 • • • • •   *c.*   4⊙ ⊙ ⊙ ⊙ •
          2 • • • • ⊙   2⊙ ⊙ ⊙ • •
          0 • • ⊙ • •   0⊙ ⊙ • • •
         −2⊙ • • • •   −2⊙ • • • •
         −4 • • • • •   −4 • • • • •
            −4 −2 0 2 4   −4 −2 0 2 4

EXERCISE SET 2

1. C,(2,2); D(1,3); G(−2,2); H(−3,1); K(−2,−2); N(1,−3); O(2,−2)
2.

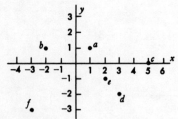

EXERCISE SET 3

1. *a.* Any 5 pairs will do as long as the *y* values are double the corresponding *x* values.
   b. $y = 2x$

c.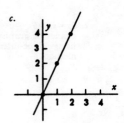

2. a. 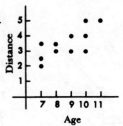 *b.* **No.**

3. a.

x	1	2	3	4
y	1	2	3	4

b. $x = y$ *c.* 1,2,3,4

4. Inside circle: *a,b,c*   Outside circle: *d,e,g,h*   On circle: *f*

5. $x^2 + y^2 = 5^2$, $x^2 + y^2 = 6^2$, $x^2 + y^2 = 7^2$, $x^2 + y^2 = r^2$

EXERCISE SET 4

1. Relations: 2,5 Functions: 1,3 **2.** *a,c* **3.** *a,d*

4. *a, b,* and *c* are the same.

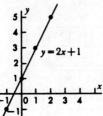

5. *a.*   *b.* 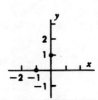   *c.*

EXERCISE SET 5

1. *a.* $y = 3x - 5$ *b.* $y = -3x - 1$ *c.* $y = 5$ *d.* $y = -\dfrac{b}{a}x + b$

2. *a.* $y = x + 1$ *b.* $y = \dfrac{1}{2}x + 5$ *c.* $y = -2x - 3$

3. *a.* $y = 2x + 4$ *b.* $y = 3x + 5$ *c.* $y = -\dfrac{2}{3}x - 2$ *d.* $y = mx - ma$

4. *a.* $y = x$ *b.* $y = x + 1$ *c.* $y = -x + 1$ *d.* $y = 2x - 4$

  e. $y = \dfrac{1}{2}x + 2$ *f.* $y = -\dfrac{1}{2}x - 1$

EXERCISE SET 6

1. *a.* slope $= 2$, *y* intercept $= -3$ *b.* slope $= \dfrac{3}{2}$, *y* intercept $= 2$

  c. slope $= -1$, *y* intercept $= 4$ *d.* slope $= \dfrac{2}{3}$, *y* intercept $= -\dfrac{5}{3}$

2.

a. 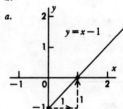 $y = x - 1$

b. 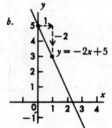 $y = -2x + 5$

c. $0y = x - \frac{1}{2}$
cannot be graphed using slope-intercept method.

3.

a. $y = \frac{1}{2}x - 3$

b. $3y - 6 = -x$

c. $4y = 0x + 8$

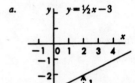
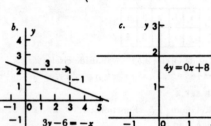

4. $F = C + 39$

F	39	50	60	70	80
C	0	11	21	31	41

5.

F	39	50	61	72	83	94
S	0	1	2	3	4	5

6. $S = 2\frac{3}{11}$ in./min.

EXERCISE SET 7

1. $P_1P_2 = 10$, $P_1P_3 = \sqrt{10}$, $P_2P_3 = \sqrt{90}$

$P_1P_3{}^2 + P_2P_3{}^2 = P_1P_2{}^2$; $10 + 90 = 100$; right angle is at P_3.

2. $P_1P_3 = P_1P_2 = \sqrt{260}$

3. $P_1P_4 = P_2P_3 = \sqrt{73}$, $P_1P_2 = P_3P_4 = \sqrt{130}$

EXERCISE SET 8

1. *a.*

x	−3	−2	−1	0	1	2	3
y	9	4	1	0	1	4	9

b.

x	−3	−2	−1	0	1	2	3
y	11	6	3	2	3	6	11

c.

x	−3	−2	−1	0	1	2	3
y	7	2	−1	−2	−1	2	7

d.

x	−3	−2	−1	0	1	2	3
y	18	8	2	0	2	8	18

e.

x	−3	−2	−1	0	1	2	3
y	9/2	2	1/2	0	1/2	2	9/2

f.

x	−3	−2	−1	0	1	2	3
y	3	0	−1	0	3	8	15

g.

x	−3	−2	−1	0	1	2	3
y	15	8	3	0	−1	0	3

h.

x	−3	−2	−1	0	1	2	3
y	4	1	0	1	4	9	16

2. *a.* $y = \frac{1}{8}x^2 + 2$

b. $y = \frac{1}{8}x^2 - \frac{1}{4}x + \frac{1}{8}$

3. $\frac{x^2}{50} + \frac{y^2}{16} = 1$ or $16x^2 + 50y^2 = 800$

1. On the sphere: *c,f*   Inside the sphere: *a,b,e,g*   Outside the sphere: *d,h*
2. $(3,0,0)$; $(0,3,0)$; $(0,0,-3)$ 3. $(3,0,0)$; $(0,-6,0)$; $(0,0,2)$
4. $x^2+y^2+z^2=6.25$

EXERCISE SET 10

1.

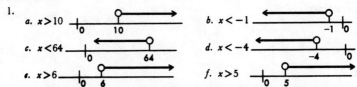

a. $x>10$ b. $x<-1$

c. $x<64$ d. $x<-4$

e. $x>6$ f. $x>5$

2.

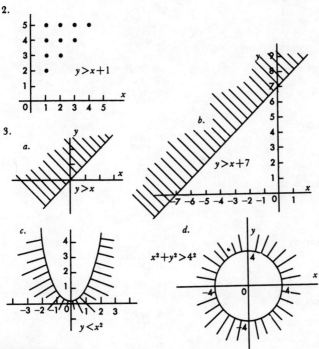

$y>x+1$

3.

a. $y>x$

b. $y>x+7$

c. $y<x^2$

d. $x^2+y^2>4^2$

EXERCISE SET 11

Students graphs will vary in scale but not in form.

EXERCISE SET 12

1. $P=x+y$; $x=2, y=3$; $P=5$ 2. $P=2x+y$; $x=2, y=3$; $P=7$
3. $C=x+y$; $x=2, y=3$; $C=5$ 4. $C=2x+y$; $x=2, y=3$; $C=7$
5. $A+B\leq60$; $A\leq2B$; $P=\$.10A+\$.12B$; $A\geq0, B\geq0$
   $A=40, B=20$; $P=\$8.80$
6. $K\geq0, M\geq0$; $.70K+.80M\geq5$; $.30K+.20M\geq2$;
   $C=K+\$.80M$; $K=6, M=1$; $C=\$6.80$

EXERCISE SET 1

1. 10^3
2. 10^5
3. 10^7
4. 10
5. 10
6. 10^2
7. 1
8. $\frac{1}{10^2} = .01$
9. $\frac{1}{10^2} = .01$
10. 1
11. 10^3
12. $\frac{1}{10^3} = .001$

EXERCISE SET 2

1. a. 230
   b. 340
   c. 4210
   d. 8020
   e. 13160
   f. 41230
   g. 16030
   h. 143.2
   i. 0.122

2. $50 \times N = \frac{N}{2} \times 100$ $500 \times N = \frac{N}{2} \times 1000$

   $.5 \times N = \frac{N}{2}$ $.05 \times N = \frac{N}{2} \div 10$

EXERCISE SET 3

1. 400
2. 1600
3. 2200
4. 1200
5. 3100

EXERCISE SET 4

1. 44.2
2. 8.84
3. 61.6
4. 12.32
5. 1.644
6. .3288

EXERCISE SET 5

1. 3872
2. 122221
3. 135795
4. 23958
5. 1086415

EXERCISE SET 6

1. 3025
2. 4225
3. 5625
4. 7225
5. 9025
6. 11025
7. 13225
8. 15625

EXERCISE SET 7

1. 3016
2. 4216
3. 5616
4. 7216
5. 9016
6. 2021
7. 3021
8. 4221
9. 5621
10. 3009
11. 1221
12. 7209
13. 3024
14. 9024
15. 7224

EXERCISE SET 8

1. 1681
2. 1764
3. 1849
4. 1936
5. 2025
6. 2209
7. 2304
8. 2401

1. 208
3. 238

2. 270
4. 361

EXERCISE SET 10
1. $(40)(40) - 1 = 1599$
3. $(60)(60) - 1 = 3599$
5. $(80)(80) - 1 = 6399$
7. $(15)(15) - 1 = 224$

2. $(50)(50) - 1 = 2499$
4. $(70)(70) - 1 = 4899$
6. $(90)(90) - 1 = 8099$
8. $(25)(25) - 1 = 624$

EXERCISE SET 11
1. 11128
3. 11881
5. 11024

2. 11124
4. 11025
6. 11009

EXERCISE SET 12
1. 9118
3. 9801
5. 9024

2. 9114
4. 9025
6. 9009

EXERCISE SET 13
1. Add 8 by adding 10 and subtracting 2.
   Add 98 by adding 100 and subtracting 2.
   Add 998 by adding 1000 and subtracting 2.
2. Add 7 by adding 10 and subtracting 3.
   Add 97 by adding 100 and subtracting 3.
   Add 997 by adding 1000 and subtracting 3.

EXERCISE SET 14
1. $8\,N = 10\,N - 2\,N$
   $98\,N = 100\,N - 2\,N$
   $998\,N = 1000\,N - 2\,N$
2. $1234 \times 90 + 51 = 1234 \times (100 - 10) + 51$
   $= 123400 - 12340 + 51$
   $= 123451 - 12340$
   $= 111111$

EXERCISE SET 15
1. Subtract 8 by subtracting 10 and adding 2.
   Subtract 98 by subtracting 100 and adding 2.
   Subtract 998 by subtracting 1000 and adding 2.
2. Subtract 7 by subtracting 10 and adding 3.
   Subtract 97 by subtracting 100 and adding 3.
   Subtract 997 by subtracting 1000 and adding 3.
3. $\$5 - \$1 + \$.05 = \$4.05$
   $\$10 - \$6 + \$.05 = \$4.05$
   $\$20 - \$17 + \$.05 = \$3.05$
   $\$20 - \$7 + \$.05 = \$13.05$

EXERCISE SET 16
1. a. $.333\ldots$
   d. $.370370\ldots$
   b. $.2424\ldots$
   e. $.141141\ldots$
   c. $.037037\ldots$
   f. $.00390039\ldots$
2. a. $\dfrac{6}{9} = \dfrac{2}{3}$
   b. $\dfrac{60}{99} = \dfrac{20}{33}$
   c. $\dfrac{600}{999} = \dfrac{200}{333}$
   d. $\dfrac{27}{99} = \dfrac{3}{11}$
   e. $\dfrac{270}{999} = \dfrac{10}{37}$
3. $352 \times .010101\ldots = 3.555\ldots$
   b. $4266 \times .001001\ldots = 4.270270\ldots$

1. *a.* $111 = 3(37)$ *b.* $1,530 = 10(153) = 10(9)(17)$
   c. $855 = 5(171) = 5(9)(19)$
   d. $4,554 = 2(2277) = 2(9)(253) = 2(9)(11)(23)$
   e. $2,088 = 8(261) = 8(9)(29)$
   f. $1,612 = 4(403) = 4(13)(31)$
2. *a.* d = 0, 2, 4, 6, or 8
   b. No value of d will work.
3. *a.* d = 1, 4, or 7
   b. d = 1, 4, or 7
4. *a.* d = 0 or 5
   b. No value of d will work.
5. *a.* d = 7
   b. d = 7
6. *a.* d = 7
   b. d = 6

1. 15810. Checking by casting out 9's gives: $4 + 2 = 6$
2. 6783. Checking by casting out 9's gives: $11 - 5 = 6$
3. 1595412. Checking by casting out 9's gives: $0 \times 2 = 0$

Solutions to the Exercises

PART V

EXERCISE SET 1

2. *a.* 1,040   *b.* 1,054.9   *c.* 2.52
   d.   587   *e.*   138   *f.* 2,108,612

3. *a.* 555   *b.* 851   *c.* 8,819
   d. 112   *e.* 6,625   *f.* 776

EXERCISE SET 2

2. *a.* 8,760   *b.* 104,775   *c.* 790,575
   d. 6,393,075   *e.* 31,977   *f.* 114,856

EXERCISE SET 3

1. $3 + 7 = 10$ **2.** $7 + 5 = 12$

$8(128) = 1024$ $(128)(32) = 4096$

3. 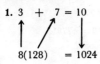 $11 - 7 = 4$ **4.** $9 - 5 = 4$

$\dfrac{2048}{128} = 16$ $\dfrac{512}{32} = 16$

5. $6 \times 2 = 12$ **6.** $4 \times 3 = 12$ **7.** $2 \times 5 = 10$
   $64^2 = 4096$ $16^3 = 4096$ $4^5 = 1024$

8. $\sqrt{4096} = \sqrt{2^{12}} = 2^{12 \div 2} = 2^6 = 64$

9. $\sqrt{256} = \sqrt{2^8} = 2^{8 \div 2} = 2^4 = 16$

10. $\sqrt[3]{4096} = \sqrt[3]{2^{12}} = 2^{12 \div 3} = 2^4 = 16$

EXERCISE SET 4

1. *a.* 2.4×10 *b.* 3.68×10^2 *c.* 4.270×10^3
   d. 2.3×10^{-1} *e.* 4.81×10^{-2} *f.* 6.42×10^0

2. *a.* $80,000 \times 5,000 = (8 \times 10^4)(5 \times 10^3) = (8 \times 5) \times 10^{4+3} = 40 \times 10^7$
   b. $20,000 \times 16,000,000 = (2 \times 10^4)(16 \times 10^6) = (2 \times 16) \times 10^{4+6} = 32 \times 10^{10}$
   c. $.0006 \times 8,000 = (6 \times 10^{-4})(8 \times 10^3) = (6 \times 8) \times 10^{-4+3} = 48 \times 10^{-1}$

   d. $.0006 \div 8,000 = \dfrac{6 \times 10^{-4}}{8 \times 10^3} = \dfrac{6}{8} \times 10^{-4-3} = .75 \times 10^{-7}$

   e. $\dfrac{4,000 \times 1,200,000}{240,000} = \dfrac{(4 \times 10^3)(12 \times 10^5)}{24 \times 10^4} = \dfrac{4 \times 12}{24} \times 10^{3+5-4} = 2 \times 10^4$

f. $\dfrac{.002 \times .005 \times .01}{.00001} = \dfrac{(2 \times 10^{-3})(5 \times 10^{-3})(1 \times 10^{-2})}{1 \times 10^{-5}} =$

$\dfrac{2 \times 5 \times 1}{1} \times 10^{-3-3-2+5} = 10 \times 10^{-3} = .01$

EXERCISE SET 5

1. *a.* $\log 82 = 1.914$ *b.* $\log 6 = .778$ *c.* $\log .07 = .845 - 2$
   d. $\log 850 = 2.929$ *e.* $\log .49 = .690 - 1$ *f.* $\log .00012 = .079 - 4$

2. *a.* 52   *b.* 3   *c.* $2.6 \times 10^3 = 2600$
   d. $1.1 \times 10^{-2} = .011$ *e.* $9.6 \times 10^{-1} = .96$ *f.* $10^{-5} = .00001$

3. *a.* $\log 5 = \log \dfrac{10}{2} = \log 10 - \log 2 = 1 - .30103 = .69897$
   b. $\log 6 = \log (2 \times 3) = \log 2 + \log 3 = .30103 + .47712 = .77815$
   c. $\log 8 = \log (2 \times 2 \times 2) = \log 2 + \log 2 + \log 2 = 3 \log 2$
      $= 3(.30103) = .90309$
   d. $\log 9 = \log (3 \times 3) = \log 3 + \log 3 = 2 \log 3 = 2(.47712) = .95424$
   e. $\log 60 = \log (2 \times 3 \times 10) = \log 2 + \log 3 + \log 10$
      $= .30103 + .47712 + 1 = 1.77815$

EXERCISE SET 6

1. *a.* $M = .903$ *b.* $M = 10^8$ *c.* $M = 1.869$
   d. $M = 10^{.740} = 5.5$ *e.* $M = 1.462$
   f. $M = 10^{.079-2} = 10^{.079} \times 10^{-2} = 1.2 \times 10^{-2} = .012$

2. *a.* $M = \sqrt{82}$

      $\log M = \dfrac{1}{2} \log 82 = \dfrac{1}{2}(1.914) = .957$

      $M = 10^{.957} = 9.05$ approximately
   b. $M = 2.3^4$
      $\log M = 4 \log 2.3 = 4(.362) = 1.448$
      $M = 10^{1.448} = 10 \times 10^{.448} = 10 \times (2.8 \text{ approximately}) = 28$

   c. $M = \dfrac{4,700 \times 85,000}{270,000}$

      $\log M = \log 4,700 + \log 85,000 - \log 270,000$
         $= 3.672 + 4.929 - 5.431 = 3.170$
      $M = 10^{3.170} = 10^3 \times 10^{.170} = 10^3 \times (1.5 \text{ approximately}) = 1,500$

   d. $M = \dfrac{.0037 \times .074}{.00016}$

      $\log M = \log .0037 + \log .074 - \log .00016$
         $= (.568 - 3) + (.869 - 2) - (.204 - 4) = .233$
      $M = 10^{.233} = 1.7$ approximately

EXERCISE SET 7

4. 4410

EXERCISE SET 8

1. $+6$	**2.** -2	**3.** $+2$	**4.** -6
5. -2	**6.** $+6$	**7.** -6	**8.** $+2$

EXERCISE SET 9

1. *a.* 8   *b.* 16   *c.* 8   *d.* .2   *e.* 2.25
     f. 25   *g.* 3   *h.* 4   *i.* 2.2 approximately   *j.* 7.1 approximately

EXERCISE SET 10

1. *a.* analog   *b.* digital   *c.* digital   *d.* analog

2. $7 = 111_{two}$

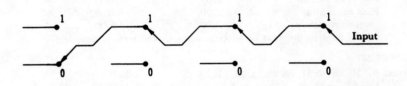

3. *a.* 10100_{two} *b.* 100000_{two} *c.* 111110_{two}

Solutions to the Exercises PART VI

EXERCISE SET 1

Answers will depend on the news, advertisements or problems in your community.

EXERCISE SET 2

1. Number of miles driven and amount of gasoline used when driving new Fords.
2. Survey the children to find the proportion who are left-handed.
3. Tabulate the reasons students are tardy during several days.
4. Have students record the time spent watching TV during at least one week.

EXERCISE SET 3

1. Duplicate subtractions. For example, Sundays, vacation and sleeping time.
2. There were many more planes flying in 1960 than in 1928.
3. Perhaps the checkered cows received special food or special treatment.
4. There may be fewer cars driven in France than Germany.
5. Most people also get over colds in seven days without using Septa.
6. The tally of earnings below shows that Plan **a** is best.

	Plan *a*	Plan *b*
First six months	$1500	$1500
Second six months	1700	1500
Third six months	1900	1750
Fourth six months	2100	1750
Fifth six months	2300	2000
Sixth six months	2500	2000

7. and 8. Analyze the news or advertisements by the method described in text.

EXERCISE SET 4

1. 36, 34, 33, 32, 31, 28, 28, 25, 23, 21, 20, 19
   Middle score is 28.
   Third from the top is 33.
   Eight scores or 67% are below 32.
3. *a.* 63-64 in.　　*b.* 74.49　　*c.* 63-64　　*d.* 20
4. *b.* 69.5-74.5,　　5,　　72
   c. 70-79,　　69.5-79.5,　　74.5
   d. 74-76,　　73.5-76.5

2. | | |
   |---|---|
   | 97 | 1 |
   | 96 | 1 |
   | 95 | 2 |
   | 94 | |
   | 93 | 3 |
   | 92 | 3 |
   | 91 | 3 |
   | 90 | 5 |
   | 89 | 1 |
   | 88 | 2 |
   | 87 | 2 |
   | 86 | 1 |

EXERCISE SET 5

1. *a.* 125 billion miles　　*b.* about 3 million
   c. 1945 to 1948　　*d.* gasoline rationing during World War II
2. *a.* transportation, clothing　　*b.* 1945-48; increased cost of production
   c. 90 cents　　*d.* $1　　*e.* $1.11　　*f.* about $13,300
3.

Score interval	Group boundaries	Frequency
29-30	28.5 — 30.5	2
27-28	26.5 — 28.5	5
25-26	24.5 — 26.5	11
23-24	22.5 — 24.5	9
21-22	20.5 — 22.5	5
19-20	18.5 — 20.5	3

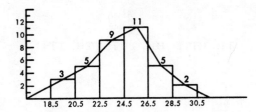

Histogram and frequency polygon

EXERCISE SET 6

1. No. We must know the total budget of each country.
2. No. The difficulty of each test is not indicated.
3. Brand B.

EXERCISE SET 7

1. *a.* 3   *b.* 21   *c.* 3   2. *a.* 4   *b.* 22.5   *c.* 3
3. *a.* 5.2   *b.* 23   *c.* 3

EXERCISE SET 8

1. *a.* 86   *b.* 21   *c.* 37   2. *a.* 85   *b.* 21   *c.* 38.2
3. *a.* 8   *b.* −.5   *c.* 11   4. *a.* 84.7   *b.* 21.2   *c.* 38.4

EXERCISE SET 9

1. Median is 24, mean is 30.1, mean is best. Median value is too low because of the widely separated groups.
2. *a.* Mode is $5, median $10, and mean $91.
   b. $5
   c. Median — it is not influenced by few extreme contributions. However, the mode is also useful in this situation, for many people would want to know the usual contribution. If it were for determining the amount a fund drive would raise, the mean would be most representative.
3. Mode, since you want the articles sold most frequently.
4. *a.* Either median (80.5) or mean (81), but median is easier to compute.
   b. Mean (79). There are two distinct groups. The median, 69.5, would put one lower score in the upper half.
5. Median (30,000), since this is not influenced by extremes (as the mean), and the distribution has 2 modes.

EXERCISE SET 10

1. *a.* fourth   *b.* 80%   *c.* 80   *d.* 9   *e.* 11   2. 79
3. *a.* 6   *b.* 18   4. *a.* fourth   *b.* 92   *c.* 14   *d.* 8   *e.* 48%
5. Barbara's rank is 25, Gary's percentile rank is 25.   6. 99, 0
7. Number of students at each school and basis of ranking.

EXERCISE SET 11

1 and 2. Statements will depend on the ads used and the city involved.
3. Each of these groups was selected according to some definite characteristic rather than at random.
4. *a.* The band members are not a random sample.
   b. Only certain businesses and companies are listed in the directory. You would get results only from people associated with certain businesses.
   c. Only a select group will mail a reply.
   d. People downtown at noon are not a random sample of the town's population.

1. $\frac{12}{30}$ or $\frac{2}{5}$ 2. *a.* $\frac{1}{4}$ *b.* $\frac{2}{4}$ or $\frac{1}{2}$

3. HHH, HHT, HTH, HTT, THH, THT, TTH, TTT

   a. $\frac{1}{8}$ *b.* $\frac{3}{8}$

4. *a.* 6   *b.* 1   *c.* $\frac{1}{6}$ *d.* $\frac{1}{6}$ *e.* $\frac{5}{6}$

5. *a.* 7   *b.* $\frac{3}{7}$ *c.* $\frac{4}{7}$ 6. $\frac{1}{52}$ 7. $\frac{13}{52}$ or $\frac{1}{4}$ 8. $\frac{4}{52}$ or $\frac{1}{13}$

9. $\frac{12}{52}$ or $\frac{3}{13}$ 10. $\frac{4}{36}$ or $\frac{1}{9}$ 11. $\frac{5}{36}$

EXERCISE SET 13

1. *a.* 2   *b.* 2.58
2. Sue   16   2
   Kay   − 4   − .5
   Charles   0   0
   Jane   12   1.5
3. Mark   10   1
   Rod   0   0
   Sue   15   1.5
   Kay   −10   −1
   Charles   5   .5
   Jane   − 5   − .5
   a. Rod   *b.* Jane   *c.* Mark

4.

Score	Frequency	*fx*	*x-M*	$(x\text{-}M)^2$	$f(x\text{-}M)^2$
0	1	0	−6	36	36
1	0	0	−5	25	0
2	1	2	−4	16	16
3	1	3	−3	9	9
4	3	12	−2	4	12
5	3	15	−1	1	3
6	5	30	0	0	0
7	8	56	1	1	8
8	2	16	2	4	8
9	2	18	3	9	18
10	1	10	4	16	16

   Mean = 6   Standard deviation = 2.16

5. $\sigma = \sqrt{\dfrac{\Sigma f(X-M)^2}{N}}$

EXERCISE SET 14

1.

Pupil	PR first test	PR second test
Mark	84	84
Rod	16	50
Sue	98	93
Kay	31	16
Charles	50	69
Jane	93	31

2.

Weight	Pounds from the mean	Standard deviations from the mean	How often expected
100	0	0	50 in 100
105	5	1	16 in 100
110	10	2	2 in 100
95	− 5	−1	84 in 100
90	−10	−2	98 in 100
102½	2½	.5	31 in 100
92½	− 7½	−1.5	93 in 100

3.

Difference	Points from mean difference	Standard deviations	How often expected
10	2	1	16%
6	−2	−1	16%
8	0	0	50%
9	1	.5	31%
5	−3	−1.5	7%
12	4	2	2%

4. *a.*

```
1   6   15   20   15   6    1
1   7   21   35   35   21   7    1
1   8   28   56   70   56   28   8   1
```

b.

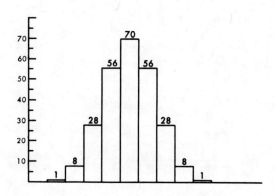

c. The distribution becomes more like the normal distribution as the number of rows is increased.

5. *a.* 7/100   *b.* 2/100   *c.* 68/100

EXERCISE SET 15

1. *a.* −   *b.* −   *c.* +   *d.* +   *e.* −   *f.* −   *g.* +
h. 0   *i.* +   *j.* +   *k.* −   *l.* +   *m.* +

2 and 3. Original illustrations.

4. *a.* Both have increased as the population has increased.
 b. Both have increased during current years of high economic growth.
 c. One has increased with the population, the other has decreased because of a polio vaccine.

A CATALOGUE OF SELECTED DOVER BOOKS
IN ALL FIELDS OF INTEREST

A CATALOGUE OF SELECTED DOVER BOOKS
IN ALL FIELDS OF INTEREST

AMERICA'S OLD MASTERS, James T. Flexner. Four men emerged unexpectedly from provincial 18th century America to leadership in European art: Benjamin West, J. S. Copley, C. R. Peale, Gilbert Stuart. Brilliant coverage of lives and contributions. Revised, 1967 edition. 69 plates. 365pp. of text.
21806-6 Paperbound $3.00

FIRST FLOWERS OF OUR WILDERNESS: AMERICAN PAINTING, THE COLONIAL PERIOD, James T. Flexner. Painters, and regional painting traditions from earliest Colonial times up to the emergence of Copley, West and Peale Sr., Foster, Gustavus Hesselius, Feke, John Smibert and many anonymous painters in the primitive manner. Engaging presentation, with 162 illustrations. xxii + 368pp.
22180-6 Paperbound $3.50

THE LIGHT OF DISTANT SKIES: AMERICAN PAINTING, 1760-1835, James T. Flexner. The great generation of early American painters goes to Europe to learn and to teach: West, Copley, Gilbert Stuart and others. Allston, Trumbull, Morse; also contemporary American painters—primitives, derivatives, academics—who remained in America. 102 illustrations. xiii + 306pp.
22179-2 Paperbound $3.50

A HISTORY OF THE RISE AND PROGRESS OF THE ARTS OF DESIGN IN THE UNITED STATES, William Dunlap. Much the richest mine of information on early American painters, sculptors, architects, engravers, miniaturists, etc. The only source of information for scores of artists, the major primary source for many others. Unabridged reprint of rare original 1834 edition, with new introduction by James T. Flexner, and 394 new illustrations. Edited by Rita Weiss. 6⅝ x 9⅝.
21695-0, 21696-9, 21697-7 Three volumes, Paperbound $15.00

EPOCHS OF CHINESE AND JAPANESE ART, Ernest F. Fenollosa. From primitive Chinese art to the 20th century, thorough history, explanation of every important art period and form, including Japanese woodcuts; main stress on China and Japan, but Tibet, Korea also included. Still unexcelled for its detailed, rich coverage of cultural background, aesthetic elements, diffusion studies, particularly of the historical period. 2nd, 1913 edition. 242 illustrations. lii + 439pp. of text.
20364-6, 20365-4 Two volumes, Paperbound $6.00

THE GENTLE ART OF MAKING ENEMIES, James A. M. Whistler. Greatest wit of his day deflates Oscar Wilde, Ruskin, Swinburne; strikes back at inane critics, exhibitions, art journalism; aesthetics of impressionist revolution in most striking form. Highly readable classic by great painter. Reproduction of edition designed by Whistler. Introduction by Alfred Werner. xxxvi + 334pp.
21875-9 Paperbound $3.00

VISUAL ILLUSIONS: THEIR CAUSES, CHARACTERISTICS, AND APPLICATIONS, Matthew Luckiesh. Thorough description and discussion of optical illusion, geometric and perspective, particularly; size and shape distortions, illusions of color, of motion; natural illusions; use of illusion in art and magic, industry, etc. Most useful today with op art, also for classical art. Scores of effects illustrated. Introduction by William H. Ittleson. 100 illustrations. xxi + 252pp.

21530-X Paperbound $2.00

A HANDBOOK OF ANATOMY FOR ART STUDENTS, Arthur Thomson. Thorough, virtually exhaustive coverage of skeletal structure, musculature, etc. Full text, supplemented by anatomical diagrams and drawings and by photographs of undraped figures. Unique in its comparison of male and female forms, pointing out differences of contour, texture, form. 211 figures, 40 drawings, 86 photographs. xx + 459pp. 5⅜ x 8⅜.

21163-0 Paperbound $3.50

150 MASTERPIECES OF DRAWING, Selected by Anthony Toney. Full page reproductions of drawings from the early 16th to the end of the 18th century, all beautifully reproduced: Rembrandt, Michelangelo, Dürer, Fragonard, Urs, Graf, Wouwerman, many others. First-rate browsing book, model book for artists. xviii + 150pp. 8⅜ x 11¼.

21032-4 Paperbound¹ $3.50

THE LATER WORK OF AUBREY BEARDSLEY, Aubrey Beardsley. Exotic, erotic, ironic masterpieces in full maturity: Comedy Ballet, Venus and Tannhauser, Pierrot, Lysistrata, Rape of the Lock, Savoy material, Ali Baba, Volpone, etc. This material revolutionized the art world, and is still powerful, fresh, brilliant. With *The Early Work*, all Beardsley's finest work. 174 plates, 2 in color. xiv + 176pp. 8⅛ x 11.

21817-1 Paperbound $3.75

DRAWINGS OF REMBRANDT, Rembrandt van Rijn. Complete reproduction of fabulously rare edition by Lippmann and Hofstede de Groot, completely reedited, updated, improved by Prof. Seymour Slive, Fogg Museum. Portraits, Biblical sketches, landscapes, Oriental types, nudes, episodes from classical mythology—All Rembrandt's fertile genius. Also selection of drawings by his pupils and followers. "Stunning volumes," *Saturday Review*. 550 illustrations. lxxviii + 552pp. 9⅛ x 12¼.

21485-0, 21486-9 Two volumes, Paperbound $10.00

THE DISASTERS OF WAR, Francisco Goya. One of the masterpieces of Western civilization—83 etchings that record Goya's shattering, bitter reaction to the Napoleonic war that swept through Spain after the insurrection of 1808 and to war in general. Reprint of the first edition, with three additional plates from Boston's Museum of Fine Arts. All plates facsimile size. Introduction by Philip Hofer, Fogg Museum. v + 97pp. 9⅜ x 8¼.

21872-4 Paperbound $2.50

GRAPHIC WORKS OF ODILON REDON. Largest collection of Redon's graphic works ever assembled: 172 lithographs, 28 etchings and engravings, 9 drawings. These include some of his most famous works. All the plates from *Odilon Redon: oeuvre graphique complet*, plus additional plates. New introduction and caption translations by Alfred Werner. 209 illustrations. xxvii + 209pp. 9⅛ x 12¼.

21966-8 Paperbound $5.00

DESIGN BY ACCIDENT; A BOOK OF "ACCIDENTAL EFFECTS" FOR ARTISTS AND DESIGNERS, James F. O'Brien. Create your own unique, striking, imaginative effects by "controlled accident" interaction of materials: paints and lacquers, oil and water based paints, splatter, crackling materials, shatter, similar items. Everything you do will be different; first book on this limitless art, so useful to both fine artist and commercial artist. Full instructions. 192 plates showing "accidents," 8 in color. viii + 215pp. 8⅜ x 11¼.   21942-9 Paperbound $3.75

THE BOOK OF SIGNS, Rudolf Koch. Famed German type designer draws 493 beautiful symbols: religious, mystical, alchemical, imperial, property marks, runes, etc. Remarkable fusion of traditional and modern. Good for suggestions of timelessness, smartness, modernity. Text. vi + 104pp. 6⅛ x 9¼.
20162-7 Paperbound $1.25

HISTORY OF INDIAN AND INDONESIAN ART, Ananda K. Coomaraswamy. An unabridged republication of one of the finest books by a great scholar in Eastern art. Rich in descriptive material, history, social backgrounds; Sunga reliefs, Rajput paintings, Gupta temples, Burmese frescoes, textiles, jewelry, sculpture, etc. 400 photos. viii + 423pp. 6⅜ x 9¾.   21436-2 Paperbound $5.00

PRIMITIVE ART, Franz Boas. America's foremost anthropologist surveys textiles, ceramics, woodcarving, basketry, metalwork, etc.; patterns, technology, creation of symbols, style origins. All areas of world, but very full on Northwest Coast Indians. More than 350 illustrations of baskets, boxes, totem poles, weapons, etc. 378 pp.
20025-6 Paperbound $3.00

THE GENTLEMAN AND CABINET MAKER'S DIRECTOR, Thomas Chippendale. Full reprint (third edition, 1762) of most influential furniture book of all time, by master cabinetmaker. 200 plates, illustrating chairs, sofas, mirrors, tables, cabinets, plus 24 photographs of surviving pieces. Biographical introduction by N. Bienenstock. vi + 249pp. 9⅞ x 12¾.   21601-2 Paperbound $4.00

AMERICAN ANTIQUE FURNITURE, Edgar G. Miller, Jr. The basic coverage of all American furniture before 1840. Individual chapters cover type of furniture—clocks, tables, sideboards, etc.—chronologically, with inexhaustible wealth of data. More than 2100 photographs, all identified, commented on. Essential to all early American collectors. Introduction by H. E. Keyes. vi + 1106pp. 7⅞ x 10¾.
21599-7, 21600-4 Two volumes, Paperbound $11.00

PENNSYLVANIA DUTCH AMERICAN FOLK ART, Henry J. Kauffman. 279 photos, 28 drawings of tulipware, Fraktur script, painted tinware, toys, flowered furniture, quilts, samplers, hex signs, house interiors, etc. Full descriptive text. Excellent for tourist, rewarding for designer, collector. Map. 146pp. 7⅞ x 10¾.
21205-X Paperbound $2.50

EARLY NEW ENGLAND GRAVESTONE RUBBINGS, Edmund V. Gillon, Jr. 43 photographs, 226 carefully reproduced rubbings show heavily symbolic, sometimes macabre early gravestones, up to early 19th century. Remarkable early American primitive art, occasionally strikingly beautiful; always powerful. Text. xxvi + 207pp. 8⅜ x 11¼.   21380-3 Paperbound $3.50

ALPHABETS AND ORNAMENTS, Ernst Lehner. Well-known pictorial source for decorative alphabets, script examples, cartouches, frames, decorative title pages, calligraphic initials, borders, similar material. 14th to 19th century, mostly European. Useful in almost any graphic arts designing, varied styles. 750 illustrations. 256pp. 7 x 10. 21905-4 Paperbound $4.00

PAINTING: A CREATIVE APPROACH, Norman Colquhoun. For the beginner simple guide provides an instructive approach to painting: major stumbling blocks for beginner; overcoming them, technical points; paints and pigments; oil painting; watercolor and other media and color. New section on "plastic" paints. Glossary. Formerly *Paint Your Own Pictures.* 221pp. 22000-1 Paperbound $1.75

THE ENJOYMENT AND USE OF COLOR, Walter Sargent. Explanation of the relations between colors themselves and between colors in nature and art, including hundreds of little-known facts about color values, intensities, effects of high and low illumination, complementary colors. Many practical hints for painters, references to great masters. 7 color plates, 29 illustrations. x + 274pp. 20944-X Paperbound $2.75

THE NOTEBOOKS OF LEONARDO DA VINCI, compiled and edited by Jean Paul Richter. 1566 extracts from original manuscripts reveal the full range of Leonardo's versatile genius: all his writings on painting, sculpture, architecture, anatomy, astronomy, geography, topography, physiology, mining, music, etc., in both Italian and English, with 186 plates of manuscript pages and more than 500 additional drawings. Includes studies for the Last Supper, the lost Sforza monument, and other works. Total of xlvii + 866pp. 7⅞ x 10¾. 22572-0, 22573-9 Two volumes, Paperbound $11.00

MONTGOMERY WARD CATALOGUE OF 1895. Tea gowns, yards of flannel and pillow-case lace, stereoscopes, books of gospel hymns, the New Improved Singer Sewing Machine, side saddles, milk skimmers, straight-edged razors, high-button shoes, spittoons, and on and on . . . listing some 25,000 items, practically all illustrated. Essential to the shoppers of the 1890's, it is our truest record of the spirit of the period. Unaltered reprint of Issue No. 57, Spring and Summer 1895. Introduction by Boris Emmet. Innumerable illustrations. xiii + 624pp. 8½ x 11⅝. 22377-9 Paperbound $6.95

THE CRYSTAL PALACE EXHIBITION ILLUSTRATED CATALOGUE (LONDON, 1851). One of the wonders of the modern world—the Crystal Palace Exhibition in which all the nations of the civilized world exhibited their achievements in the arts and sciences—presented in an equally important illustrated catalogue. More than 1700 items pictured with accompanying text—ceramics, textiles, cast-iron work, carpets, pianos, sleds, razors, wall-papers, billiard tables, beehives, silverware and hundreds of other artifacts—represent the focal point of Victorian culture in the Western World. Probably the largest collection of Victorian decorative art ever assembled—indispensable for antiquarians and designers. Unabridged republication of the Art-Journal Catalogue of the Great Exhibition of 1851, with all terminal essays. New introduction by John Gloag, F.S.A. xxxiv + 426pp. 9 x 12. 22503-8 Paperbound $5.00

A HISTORY OF COSTUME, Carl Köhler. Definitive history, based on surviving pieces of clothing primarily, and paintings, statues, etc. secondarily. Highly readable text, supplemented by 594 illustrations of costumes of the ancient Mediterranean peoples, Greece and Rome, the Teutonic prehistoric period; costumes of the Middle Ages, Renaissance, Baroque, 18th and 19th centuries. Clear, measured patterns are provided for many clothing articles. Approach is practical throughout. Enlarged by Emma von Sichart. 464pp.   21030-8 Paperbound $3.50.

ORIENTAL RUGS, ANTIQUE AND MODERN, Walter A. Hawley. A complete and authoritative treatise on the Oriental rug—where they are made, by whom and how, designs and symbols, characteristics in detail of the six major groups, how to distinguish them and how to buy them. Detailed technical data is provided on periods, weaves, warps, wefts, textures, sides, ends and knots, although no technical background is required for an understanding. 11 color plates, 80 halftones, 4 maps. vi + 320pp. 6⅛ x 9⅛.   22366-3 Paperbound $5.00

TEN BOOKS ON ARCHITECTURE, Vitruvius. By any standards the most important book on architecture ever written. Early Roman discussion of aesthetics of building, construction methods, orders, sites, and every other aspect of architecture has inspired, instructed architecture for about 2,000 years. Stands behind Palladio, Michelangelo, Bramante, Wren, countless others. Definitive Morris H. Morgan translation. 68 illustrations. xii + 331pp.   20645-9 Paperbound $3.00

THE FOUR BOOKS OF ARCHITECTURE, Andrea Palladio. Translated into every major Western European language in the two centuries following its publication in 1570, this has been one of the most influential books in the history of architecture. Complete reprint of the 1738 Isaac Ware edition. New introduction by Adolf Placzek, Columbia Univ. 216 plates. xxii + 110pp. of text. 9½ x 12¾.   21308-0 Clothbound $12.50

STICKS AND STONES: A STUDY OF AMERICAN ARCHITECTURE AND CIVILIZATION, Lewis Mumford.One of the great classics of American cultural history. American architecture from the medieval-inspired earliest forms to the early 20th century; evolution of structure and style, and reciprocal influences on environment. 21 photographic illustrations. 238pp.   20202-X Paperbound $2.00

THE AMERICAN BUILDER'S COMPANION, Asher Benjamin. The most widely used early 19th century architectural style and source book, for colonial up into Greek Revival periods. Extensive development of geometry of carpentering, construction of sashes, frames, doors, stairs; plans and elevations of domestic and other buildings. Hundreds of thousands of houses were built according to this book, now invaluable to historians, architects, restorers, etc. 1827 edition. 59 plates. 114pp. 7⅞ x 10¾.   22236-5 Paperbound $3.50

DUTCH HOUSES IN THE HUDSON VALLEY BEFORE 1776, Helen Wilkinson Reynolds. The standard survey of the Dutch colonial house and outbuildings, with constructional features, decoration, and local history associated with individual homesteads. Introduction by Franklin D. Roosevelt. Map. 150 illustrations. 469pp. 6⅝ x 9¼.   21469-9 Paperbound $5.00

THE ARCHITECTURE OF COUNTRY HOUSES, Andrew J. Downing. Together with Vaux's *Villas and Cottages* this is the basic book for Hudson River Gothic architecture of the middle Victorian period. Full, sound discussions of general aspects of housing, architecture, style, decoration, furnishing, together with scores of detailed house plans, illustrations of specific buildings, accompanied by full text. Perhaps the most influential single American architectural book. 1850 edition. Introduction by J. Stewart Johnson. 321 figures, 34 architectural designs. xvi + 560pp.
22003-6 Paperbound $4.00

LOST EXAMPLES OF COLONIAL ARCHITECTURE, John Mead Howells. Full-page photographs of buildings that have disappeared or been so altered as to be denatured, including many designed by major early American architects. 245 plates. xvii + 248pp. 7⅞ x 10¾. 21143-6 Paperbound $3.50

DOMESTIC ARCHITECTURE OF THE AMERICAN COLONIES AND OF THE EARLY REPUBLIC, Fiske Kimball. Foremost architect and restorer of Williamsburg and Monticello covers nearly 200 homes between 1620-1825. Architectural details, construction, style features, special fixtures, floor plans, etc. Generally considered finest work in its area. 219 illustrations of houses, doorways, windows, capital mantels. xx + 314pp. 7⅞ x 10¾. 21743-4 Paperbound $4.00

EARLY AMERICAN ROOMS: 1650-1858, edited by Russell Hawes Kettell. Tour of 12 rooms, each representative of a different era in American history and each furnished, decorated, designed and occupied in the style of the era. 72 plans and elevations, 8-page color section, etc., show fabrics, wall papers, arrangements, etc. Full descriptive text. xvii + 200pp. of text. 8⅜ x 11¼.
21633-0 Paperbound $5.00

THE FITZWILLIAM VIRGINAL BOOK, edited by J. Fuller Maitland and W. B. Squire. Full modern printing of famous early 17th-century ms. volume of 300 works by Morley, Byrd, Bull, Gibbons, etc. For piano or other modern keyboard instrument; easy to read format. xxxvi + 938pp. 8⅜ x 11.
21068-5, 21069-3 Two volumes, Paperbound $10.00

KEYBOARD MUSIC, Johann Sebastian Bach. Bach Gesellschaft edition. A rich selection of Bach's masterpieces for the harpsichord: the six English Suites, six French Suites, the six Partitas (Clavierübung part I), the Goldberg Variations (Clavierübung part IV), the fifteen Two-Part Inventions and the fifteen Three-Part Sinfonias. Clearly reproduced on large sheets with ample margins; eminently playable. vi + 312pp. 8⅛ x 11. 22360-4 Paperbound $5.00

THE MUSIC OF BACH: AN INTRODUCTION, Charles Sanford Terry. A fine, nontechnical introduction to Bach's music, both instrumental and vocal. Covers organ music, chamber music, passion music, other types. Analyzes themes, developments, innovations. x + 114pp. 21075-8 Paperbound $1.50

BEETHOVEN AND HIS NINE SYMPHONIES, Sir George Grove. Noted British musicologist provides best history, analysis, commentary on symphonies. Very thorough, rigorously accurate; necessary to both advanced student and amateur music lover. 436 musical passages. vii + 407 pp. 20334-4 Paperbound $2.75

JOHANN SEBASTIAN BACH, Philipp Spitta. One of the great classics of musicology, this definitive analysis of Bach's music (and life) has never been surpassed. Lucid, nontechnical analyses of hundreds of pieces (30 pages devoted to St. Matthew Passion, 26 to B Minor Mass). Also includes major analysis of 18th-century music. 450 musical examples. 40-page musical supplement. Total of xx + 1799pp.

(EUK) 22278-0, 22279-9 Two volumes, Clothbound $17.50

MOZART AND HIS PIANO CONCERTOS, Cuthbert Girdlestone. The only full-length study of an important area of Mozart's creativity. Provides detailed analyses of all 23 concertos, traces inspirational sources. 417 musical examples. Second edition. 509pp. 21271-8 Paperbound $3.50

THE PERFECT WAGNERITE: A COMMENTARY ON THE NIBLUNG'S RING, George Bernard Shaw. Brilliant and still relevant criticism in remarkable essays on Wagner's Ring cycle, Shaw's ideas on political and social ideology behind the plots, role of Leitmotifs, vocal requisites, etc. Prefaces. xxi + 136pp.

(USO) 21707-8 Paperbound $1.75

DON GIOVANNI, W. A. Mozart. Complete libretto, modern English translation; biographies of composer and librettist; accounts of early performances and critical reaction. Lavishly illustrated. All the material you need to understand and appreciate this great work. Dover Opera Guide and Libretto Series; translated and introduced by Ellen Bleiler. 92 illustrations. 209pp.

21134-7 Paperbound $2.00

BASIC ELECTRICITY, U. S. Bureau of Naval Personel. Originally a training course, best non-technical coverage of basic theory of electricity and its applications. Fundamental concepts, batteries, circuits, conductors and wiring techniques, AC and DC, inductance and capacitance, generators, motors, transformers, magnetic amplifiers, synchros, servomechanisms, etc. Also covers blue-prints, electrical diagrams, etc. Many questions, with answers. 349 illustrations. x + 448pp. 6½ x 9¼.

20973-3 Paperbound $3.50

REPRODUCTION OF SOUND, Edgar Villchur. Thorough coverage for laymen of high fidelity systems, reproducing systems in general, needles, amplifiers, preamps, loudspeakers, feedback, explaining physical background. "A rare talent for making technicalities vividly comprehensible," R. Darrell, *High Fidelity.* 69 figures. iv + 92pp. 21515-6 Paperbound $1.35

HEAR ME TALKIN' TO YA: THE STORY OF JAZZ AS TOLD BY THE MEN WHO MADE IT, Nat Shapiro and Nat Hentoff. Louis Armstrong, Fats Waller, Jo Jones, Clarence Williams, Billy Holiday, Duke Ellington, Jelly Roll Morton and dozens of other jazz greats tell how it was in Chicago's South Side, New Orleans, depression Harlem and the modern West Coast as jazz was born and grew. xvi + 429pp.

21726-4 Paperbound $3.00

FABLES OF AESOP, translated by Sir Roger L'Estrange. A reproduction of the very rare 1931 Paris edition; a selection of the most interesting fables, together with 50 imaginative drawings by Alexander Calder. v + 128pp. 6½x9¼.

21780-9 Paperbound $1.50

AGAINST THE GRAIN (A REBOURS), Joris K. Huysmans. Filled with weird images, evidences of a bizarre imagination, exotic experiments with hallucinatory drugs, rich tastes and smells and the diversions of its sybarite hero Duc Jean des Esseintes, this classic novel pushed 19th-century literary decadence to its limits. Full unabridged edition. Do not confuse this with abridged editions generally sold. Introduction by Havelock Ellis. xlix + 206pp.   22190-3 Paperbound $2.50

VARIORUM SHAKESPEARE: HAMLET. Edited by Horace H. Furness; a landmark of American scholarship. Exhaustive footnotes and appendices treat all doubtful words and phrases, as well as suggested critical emendations throughout the play's history. First volume contains editor's own text, collated with all Quartos and Folios. Second volume contains full first Quarto, translations of Shakespeare's sources (Belleforest, and Saxo Grammaticus), Der Bestrafte Brudermord, and many essays on critical and historical points of interest by major authorities of past and present. Includes details of staging and costuming over the years. By far the best edition available for serious students of Shakespeare. Total of xx + 905pp.
21004-9, 21005-7, 2 volumes, Paperbound $7.00

A LIFE OF WILLIAM SHAKESPEARE, Sir Sidney Lee. This is the standard life of Shakespeare, summarizing everything known about Shakespeare and his plays. Incredibly rich in material, broad in coverage, clear and judicious, it has served thousands as the best introduction to Shakespeare. 1931 edition. 9 plates. xxix + 792pp.   21967-4 Paperbound $4.50

MASTERS OF THE DRAMA, John Gassner. Most comprehensive history of the drama in print, covering every tradition from Greeks to modern Europe and America, including India, Far East, etc. Covers more than 800 dramatists, 2000 plays, with biographical material, plot summaries, theatre history, criticism, etc. "Best of its kind in English," *New Republic*. 77 illustrations. xxii + 890pp.
20100-7 Clothbound $10.00

THE EVOLUTION OF THE ENGLISH LANGUAGE, George McKnight. The growth of English, from the 14th century to the present. Unusual, non-technical account presents basic information in very interesting form: sound shifts, change in grammar and syntax, vocabulary growth, similar topics. Abundantly illustrated with quotations. Formerly *Modern English in the Making*. xii + 590pp.
21932-1 Paperbound $4.00

AN ETYMOLOGICAL DICTIONARY OF MODERN ENGLISH, Ernest Weekley. Fullest, richest work of its sort, by foremost British lexicographer. Detailed word histories, including many colloquial and archaic words; extensive quotations. Do not confuse this with the Concise Etymological Dictionary, which is much abridged. Total of xxvii + 830pp. 6½ x 9¼.
21873-2, 21874-0 Two volumes, Paperbound $7.90

FLATLAND: A ROMANCE OF MANY DIMENSIONS, E. A. Abbott. Classic of science-fiction explores ramifications of life in a two-dimensional world, and what happens when a three-dimensional being intrudes. Amusing reading, but also useful as introduction to thought about hyperspace. Introduction by Banesh Hoffmann. 16 illustrations. xx + 103pp.   20001-9 Paperbound $1.25

POEMS OF ANNE BRADSTREET, edited with an introduction by Robert Hutchinson. A new selection of poems by America's first poet and perhaps the first significant woman poet in the English language. 48 poems display her development in works of considerable variety—love poems, domestic poems, religious meditations, formal elegies, "quaternions," etc. Notes, bibliography. viii + 222pp.

22160-1 Paperbound $2.50

THREE GOTHIC NOVELS: THE CASTLE OF OTRANTO BY HORACE WALPOLE; VATHEK BY WILLIAM BECKFORD; THE VAMPYRE BY JOHN POLIDORI, WITH FRAGMENT OF A NOVEL BY LORD BYRON, edited by E. F. Bleiler. The first Gothic novel, by Walpole; the finest Oriental tale in English, by Beckford; powerful Romantic supernatural story in versions by Polidori and Byron. All extremely important in history of literature; all still exciting, packed with supernatural thrills, ghosts, haunted castles, magic, etc. xl + 291pp.

21232-7 Paperbound $2.50

THE BEST TALES OF HOFFMANN, E. T. A. Hoffmann. 10 of Hoffmann's most important stories, in modern re-editings of standard translations: Nutcracker and the King of Mice, Signor Formica, Automata, The Sandman, Rath Krespel, The Golden Flowerpot, Master Martin the Cooper, The Mines of Falun, The King's Betrothed, A New Year's Eve Adventure. 7 illustrations by Hoffmann. Edited by E. F. Bleiler. xxxix + 419pp.

21793-0 Paperbound $3.00

GHOST AND HORROR STORIES OF AMBROSE BIERCE, Ambrose Bierce. 23 strikingly modern stories of the horrors latent in the human mind: The Eyes of the Panther, The Damned Thing, An Occurrence at Owl Creek Bridge, An Inhabitant of Carcosa, etc., plus the dream-essay, Visions of the Night. Edited by E. F. Bleiler. xxii + 199pp.

20767-6 Paperbound $1.50

BEST GHOST STORIES OF J. S. LeFANU, J. Sheridan LeFanu. Finest stories by Victorian master often considered greatest supernatural writer of all. Carmilla, Green Tea, The Haunted Baronet, The Familiar, and 12 others. Most never before available in the U. S. A. Edited by E. F. Bleiler. 8 illustrations from Victorian publications. xvii + 467pp.

20415-4 Paperbound $3.00

MATHEMATICAL FOUNDATIONS OF INFORMATION THEORY, A. I. Khinchin. Comprehensive introduction to work of Shannon, McMillan, Feinstein and Khinchin, placing these investigations on a rigorous mathematical basis. Covers entropy concept in probability theory, uniqueness theorem, Shannon's inequality, ergodic sources, the E property, martingale concept, noise, Feinstein's fundamental lemma, Shanon's first and second theorems. Translated by R. A. Silverman and M. D. Friedman. iii + 120pp.

60434-9 Paperbound $2.00

SEVEN SCIENCE FICTION NOVELS, H. G. Wells. The standard collection of the great novels. Complete, unabridged. *First Men in the Moon, Island of Dr. Moreau, War of the Worlds, Food of the Gods, Invisible Man, Time Machine, In the Days of the Comet.* Not only science fiction fans, but every educated person owes it to himself to read these novels. 1015pp. (USO) 20264-X Clothbound $6.00

LAST AND FIRST MEN AND STAR MAKER, TWO SCIENCE FICTION NOVELS, Olaf Stapledon. Greatest future histories in science fiction. In the first, human intelligence is the "hero," through strange paths of evolution, interplanetary invasions, incredible technologies, near extinctions and reemergences. Star Maker describes the quest of a band of star rovers for intelligence through time and space: weird inhuman civilizations, crustacean minds, symbiotic worlds, etc. Complete, unabridged. v + 438pp. (USO) 21962-3 Paperbound $2.50

THREE PROPHETIC NOVELS, H. G. WELLS. Stages of a consistently planned future for mankind. *When the Sleeper Wakes,* and *A Story of the Days to Come,* anticipate *Brave New World* and *1984,* in the 21st Century; *The Time Machine,* only complete version in print, shows farther future and the end of mankind. All show Wells's greatest gifts as storyteller and novelist. Edited by E. F. Bleiler. x + 335pp. (USO) 20605-X Paperbound $2.50

THE DEVIL'S DICTIONARY, Ambrose Bierce. America's own Oscar Wilde— Ambrose Bierce—offers his barbed iconoclastic wisdom in over 1,000 definitions hailed by H. L. Mencken as "some of the most gorgeous witticisms in the English language." 145pp. 20487-1 Paperbound $1.25

MAX AND MORITZ, Wilhelm Busch. Great children's classic, father of comic strip, of two bad boys, Max and Moritz. Also Ker and Plunk (Plisch und Plumm), Cat and Mouse, Deceitful Henry, Ice-Peter, The Boy and the Pipe, and five other pieces. Original German, with English translation. Edited by H. Arthur Klein; translations by various hands and H. Arthur Klein. vi + 216pp. 20181-3 Paperbound $2.00

PIGS IS PIGS AND OTHER FAVORITES, Ellis Parker Butler. The title story is one of the best humor short stories, as Mike Flannery obfuscates biology and English. Also included, That Pup of Murchison's, The Great American Pie Company, and Perkins of Portland. 14 illustrations. v + 109pp. 21532-6 Paperbound $1.25

THE PETERKIN PAPERS, Lucretia P. Hale. It takes genius to be as stupidly mad as the Peterkins, as they decide to become wise, celebrate the "Fourth," keep a cow, and otherwise strain the resources of the Lady from Philadelphia. Basic book of American humor. 153 illustrations. 219pp. 20794-3 Paperbound $2.00

PERRAULT'S FAIRY TALES, translated by A. E. Johnson and S. R. Littlewood, with 34 full-page illustrations by Gustave Doré. All the original Perrault stories— Cinderella, Sleeping Beauty, Bluebeard, Little Red Riding Hood, Puss in Boots, Tom Thumb, etc.—with their witty verse morals and the magnificent illustrations of Doré. One of the five or six great books of European fairy tales. viii + 117pp. 8⅛ x 11. 22311-6 Paperbound $2.00

OLD HUNGARIAN FAIRY TALES, Baroness Orczy. Favorites translated and adapted by author of the *Scarlet Pimpernel.* Eight fairy tales include "The Suitors of Princess Fire-Fly," "The Twin Hunchbacks," "Mr. Cuttlefish's Love Story," and "The Enchanted Cat." This little volume of magic and adventure will captivate children as it has for generations. 90 drawings by Montagu Barstow. 96pp. (USO) 22293-4 Paperbound $1.95

THE RED FAIRY BOOK, Andrew Lang. Lang's color fairy books have long been children's favorites. This volume includes Rapunzel, Jack and the Bean-stalk and 35 other stories, familiar and unfamiliar. 4 plates, 93 illustrations x + 367pp.
21673-X Paperbound $2.50

THE BLUE FAIRY BOOK, Andrew Lang. Lang's tales come from all countries and all times. Here are 37 tales from Grimm, the Arabian Nights, Greek Mythology, and other fascinating sources. 8 plates, 130 illustrations. xi + 390pp.
21437-0 Paperbound $2.75

HOUSEHOLD STORIES BY THE BROTHERS GRIMM. Classic English-language edition of the well-known tales — Rumpelstiltskin, Snow White, Hansel and Gretel, The Twelve Brothers, Faithful John, Rapunzel, Tom Thumb (52 stories in all). Translated into simple, straightforward English by Lucy Crane. Ornamented with headpieces, vignettes, elaborate decorative initials and a dozen full-page illustrations by Walter Crane. x + 269pp. 21080-4 Paperbound **$2.00**

THE MERRY ADVENTURES OF ROBIN HOOD, Howard Pyle. The finest modern versions of the traditional ballads and tales about the great English outlaw. Howard Pyle's complete prose version, with every word, every illustration of the first edition. Do not confuse this facsimile of the original (1883) with modern editions that change text or illustrations. 23 plates plus many page decorations. xxii + 296pp.
22043-5 Paperbound $2.75

THE STORY OF KING ARTHUR AND HIS KNIGHTS, Howard Pyle. The finest children's version of the life of King Arthur; brilliantly retold by Pyle, with 48 of his most imaginative illustrations. xviii + 313pp. 6⅛ x 9¼.
21445-1 Paperbound $2.50

THE WONDERFUL WIZARD OF OZ, L. Frank Baum. America's finest children's book in facsimile of first edition with all Denslow illustrations in full color. The edition a child should have. Introduction by Martin Gardner. 23 color plates, scores of drawings. iv + 267pp. 20691-2 Paperbound $2.50

THE MARVELOUS LAND OF OZ, L. Frank Baum. The second Oz book, every bit as imaginative as the Wizard. The hero is a boy named Tip, but the Scarecrow and the Tin Woodman are back, as is the Oz magic. 16 color plates, 120 drawings by John R. Neill. 287pp. 20692-0 Paperbound $2.50

THE MAGICAL MONARCH OF MO, L. Frank Baum. Remarkable adventures in a land even stranger than Oz. The best of Baum's books not in the Oz series. 15 color plates and dozens of drawings by Frank Verbeck. xviii + 237pp.
21892-9 Paperbound $2.25

THE BAD CHILD'S BOOK OF BEASTS, MORE BEASTS FOR WORSE CHILDREN, A MORAL ALPHABET, Hilaire Belloc. Three complete humor classics in one volume. Be kind to the frog, and do not call him names . . . and 28 other whimsical animals. Familiar favorites and some not so well known. Illustrated by Basil Blackwell. 156pp. (USO) 20749-8 Paperbound $1.50

EAST O' THE SUN AND WEST O' THE MOON, George W. Dasent. Considered the best of all translations of these Norwegian folk tales, this collection has been enjoyed by generations of children (and folklorists too). Includes True and Untrue, Why the Sea is Salt, East O' the Sun and West O' the Moon, Why the Bear is Stumpy-Tailed, Boots and the Troll, The Cock and the Hen, Rich Peter the Pedlar, and 52 more. The only edition with all 59 tales. 77 illustrations by Erik Werenskiold and Theodor Kittelsen. xv + 418pp. 22521-6 Paperbound $3.50

GOOPS AND HOW TO BE THEM, Gelett Burgess. Classic of tongue-in-cheek humor, masquerading as etiquette book. 87 verses, twice as many cartoons, show mischievous Goops as they demonstrate to children virtues of table manners, neatness, courtesy, etc. Favorite for generations. viii + 88pp. 6½ x 9¼.
22233-0 Paperbound $1.50

ALICE'S ADVENTURES UNDER GROUND, Lewis Carroll. The first version, quite different from the final *Alice in Wonderland,* printed out by Carroll himself with his own illustrations. Complete facsimile of the "million dollar" manuscript Carroll gave to Alice Liddell in 1864. Introduction by Martin Gardner. viii + 96pp. Title and dedication pages in color. 21482-6 Paperbound $1.25

THE BROWNIES, THEIR BOOK, Palmer Cox. Small as mice, cunning as foxes, exuberant and full of mischief, the Brownies go to the zoo, toy shop, seashore, circus, etc., in 24 verse adventures and 266 illustrations. Long a favorite, since their first appearance in St. Nicholas Magazine. xi + 144pp. 6⅝ x 9¼.
21265-3 Paperbound $1.75

SONGS OF CHILDHOOD, Walter De La Mare. Published (under the pseudonym Walter Ramal) when De La Mare was only 29, this charming collection has long been a favorite children's book. A facsimile of the first edition in paper, the 47 poems capture the simplicity of the nursery rhyme and the ballad, including such lyrics as I Met Eve, Tartary, The Silver Penny. vii + 106pp. (USO) 21972-0 Paperbound $2.00

THE COMPLETE NONSENSE OF EDWARD LEAR, Edward Lear. The finest 19th-century humorist-cartoonist in full: all nonsense limericks, zany alphabets, Owl and Pussycat, songs, nonsense botany, and more than 500 illustrations by Lear himself. Edited by Holbrook Jackson. xxix + 287pp. (USO) 20167-8 Paperbound $2.00

BILLY WHISKERS: THE AUTOBIOGRAPHY OF A GOAT, Frances Trego Montgomery. A favorite of children since the early 20th century, here are the escapades of that rambunctious, irresistible and mischievous goat—Billy Whiskers. Much in the spirit of *Peck's Bad Boy,* this is a book that children never tire of reading or hearing. All the original familiar illustrations by W. H. Fry are included: 6 color plates, 18 black and white drawings. 159pp. 22345-0 Paperbound $2.00

MOTHER GOOSE MELODIES. Faithful republication of the fabulously rare Munroe and Francis "copyright 1833" Boston edition—the most important Mother Goose collection, usually referred to as the "original." Familiar rhymes plus many rare ones, with wonderful old woodcut illustrations. Edited by E. F. Bleiler. 128pp. 4½ x 6⅜. 22577-1 Paperbound $1.00

Two Little Savages; Being the Adventures of Two Boys Who Lived as Indians and What They Learned, Ernest Thompson Seton. Great classic of nature and boyhood provides a vast range of woodlore in most palatable form, a genuinely entertaining story. Two farm boys build a teepee in woods and live in it for a month, working out Indian solutions to living problems, star lore, birds and animals, plants, etc. 293 illustrations. vii + 286pp.

20985-7 Paperbound $2.50

Peter Piper's Practical Principles of Plain & Perfect Pronunciation. Alliterative jingles and tongue-twisters of surprising charm, that made their first appearance in America about 1830. Republished in full with the spirited woodcut illustrations from this earliest American edition. 32pp. 4½ x 6⅜.

22560-7 Paperbound $1.00

Science Experiments and Amusements for Children, Charles Vivian. 73 easy experiments, requiring only materials found at home or easily available, such as candles, coins, steel wool, etc.; illustrate basic phenomena like vacuum, simple chemical reaction, etc. All safe. Modern, well-planned. Formerly *Science Games for Children.* 102 photos, numerous drawings. 96pp. 6⅛ x 9¼.

21856-2 Paperbound $1.25

An Introduction to Chess Moves and Tactics Simply Explained, Leonard Barden. Informal intermediate introduction, quite strong in explaining reasons for moves. Covers basic material, tactics, important openings, traps, positional play in middle game, end game. Attempts to isolate patterns and recurrent configurations. Formerly *Chess.* 58 figures. 102pp. (USO) 21210-6 Paperbound $1.25

Lasker's Manual of Chess, Dr. Emanuel Lasker. Lasker was not only one of the five great World Champions, he was also one of the ablest expositors, theorists, and analysts. In many ways, his Manual, permeated with his philosophy of battle, filled with keen insights, is one of the greatest works ever written on chess. Filled with analyzed games by the great players. A single-volume library that will profit almost any chess player, beginner or master. 308 diagrams. xli x 349pp.

20640-8 Paperbound $2.75

The Master Book of Mathematical Recreations, Fred Schuh. In opinion of many the finest work ever prepared on mathematical puzzles, stunts, recreations; exhaustively thorough explanations of mathematics involved, analysis of effects, citation of puzzles and games. Mathematics involved is elementary. Translated bv F. Göbel. 194 figures. xxiv + 430pp.

22134-2 Paperbound $3.50

Mathematics, Magic and Mystery, Martin Gardner. Puzzle editor for Scientific American explains mathematics behind various mystifying tricks: card tricks, stage "mind reading," coin and match tricks, counting out games, geometric dissections, etc. Probability sets, theory of numbers clearly explained. Also provides more than 400 tricks, guaranteed to work, that you can do. 135 illustrations. xii + 176pp.

20335-2 Paperbound $1.75

MATHEMATICAL PUZZLES FOR BEGINNERS AND ENTHUSIASTS, Geoffrey Mott-Smith. 189 puzzles from easy to difficult—involving arithmetic, logic, algebra, properties of digits, probability, etc.—for enjoyment and mental stimulus. Explanation of mathematical principles behind the puzzles. 135 illustrations. viii + 248pp.

20198-8 Paperbound $1.75

PAPER FOLDING FOR BEGINNERS, William D. Murray and Francis J. Rigney. Easiest book on the market, clearest instructions on making interesting, beautiful origami. Sail boats, cups, roosters, frogs that move legs, bonbon boxes, standing birds, etc. 40 projects; more than 275 diagrams and photographs. 94pp.

20713-7 Paperbound $1.00

TRICKS AND GAMES ON THE POOL TABLE, Fred Herrmann. 79 tricks and games— some solitaires, some for two or more players, some competitive games—to entertain you between formal games. Mystifying shots and throws, unusual caroms, tricks involving such props as cork, coins, a hat, etc. Formerly *Fun on the Pool Table*. 77 figures. 95pp.

21814-7 Paperbound $1.25

HAND SHADOWS TO BE THROWN UPON THE WALL: A SERIES OF NOVEL AND AMUSING FIGURES FORMED BY THE HAND, Henry Bursill. Delightful picturebook from great-grandfather's day shows how to make 18 different hand shadows: a bird that flies, duck that quacks, dog that wags his tail, camel, goose, deer, boy, turtle, etc. Only book of its sort. vi + 33pp. 6½ x 9¼.

21779-5 Paperbound $1.00

WHITTLING AND WOODCARVING, E. J. Tangerman. 18th printing of best book on market. "If you can cut a potato you can carve" toys and puzzles, chains, chessmen, caricatures, masks, frames, woodcut blocks, surface patterns, much more. Information on tools, woods, techniques. Also goes into serious wood sculpture from Middle Ages to present, East and West. 464 photos, figures. x + 293pp.

20965-2 Paperbound $2.00

HISTORY OF PHILOSOPHY, Julián Marias. Possibly the clearest, most easily followed, best planned, most useful one-volume history of philosophy on the market; neither skimpy nor overfull. Full details on system of every major philosopher and dozens of less important thinkers from pre-Socratics up to Existentialism and later. Strong on many European figures usually omitted. Has gone through dozens of editions in Europe. 1966 edition, translated by Stanley Appelbaum and Clarence Strowbridge. xviii + 505pp.

21739-6 Paperbound $3.50

YOGA: A SCIENTIFIC EVALUATION, Kovoor T. Behanan. Scientific but non-technical study of physiological results of yoga exercises; done under auspices of Yale U. Relations to Indian thought, to psychoanalysis, etc. 16 photos. xxiii + 270pp.

20505-3 Paperbound $2.50

Prices subject to change without notice.

Available at your book dealer or write for free catalogue to Dept. GI, Dover Publications, Inc., 180 Varick St., N. Y., N. Y. 10014. Dover publishes more than 150 books each year on science, elementary and advanced mathematics, biology, music, art, literary history, social sciences and other areas.